# Use R!

*Series Editors:*
Robert Gentleman   Kurt Hornik   Giovanni Parmigiani

For other titles published in this series, go to
http://www.springer.com/series/6991

Andrea S. Foulkes

# Applied Statistical Genetics with R

## For Population-based Association Studies

 Springer

Andrea S. Foulkes
University of Massachusetts
School of Public Health & Health Sciences
404 Arnold House
715 N. Pleasant Street
Amherst, MA 01003
USA
foulkes@schoolph.umass.edu

ISBN 978-0-387-89553-6      e-ISBN 978-0-387-89554-3
DOI 10.1007/978-0-387-89554-3
Springer Dordrecht Heidelberg London New York

Library of Congress Control Number: PCN applied for

Printed on acid-free paper

Springer is part of Springer Science+Business Media (www.springer.com)

To Rich, Sophie and Ella

# Preface

This book is intended to provide fundamental statistical concepts and tools relevant to the analysis of genetic data arising from population-based association studies. Elementary knowledge of statistical methods at the level of a first course in biostatistics is assumed. Chapters 1–3 provide a general overview of the genetic and epidemiological considerations relevant to this setting. Topics covered include: (1) types of investigations, typical data components and features in genetic association studies, and basic genetic vocabulary (Chapter 1); (2) epidemiological principles relevant to population-based studies, including confounding and effect modification (Chapter 2); (3) elementary statistical methods for estimating and testing association (Chapter 2); (4) the overarching analytical challenges inherent in these investigations (Chapter 2); (5) basic genetic concepts, including linkage disequilibrium, Hardy-Weinberg equilibrium, and haplotypic phase (Chapter 3); and (6) quality control methods for assessing genotyping errors and population substructure (Chapter 3).

The remaining chapters are organized as follows. Chapters 4 and 5 deal primarily with methods that aim to identify single genetic polymorphisms or single genes that contribute individually to measures of disease progression or disease status. This includes testing concepts and methods for appropriately adjusting for multiple comparisons (Chapter 4) and approaches to the analysis of unobservable haplotypic phase (Chapter 5). Chapters 6 and 7 focus on methods for variable subset selection and particularly methods that simultaneously evaluate a large number of variables to arrive at the best predictive model for the complex disease trait under investigation. Notably, while all of these methods consider multiple polymorphisms concomitantly, some focus on conditional effects of these genetic variables, while other methods are specifically designed for identifying and testing potential interaction among genetic polymorphisms in their effects on disease phenotypes. This section covers classification and regression trees (Chapter 6), extensions of the tree framework—namely random forests, logic regression and multivariable adaptive regression splines—and a brief introduction to Bayesian variable selection (Chapter 7).

The field of statistical genomics includes a large array of methods for a wide variety of medical and public health applications. While the methods described herein are broadly relevant, this text does not directly address issues specific to family-based studies, evolutionary (population genetic) modeling, and gene expression analysis. This text also does not attempt to provide a comprehensive summary of existing methods in the rapidly expanding field of statistical genomics. Rather, fundamental concepts are presented at the level of an introductory graduate-level course in biostatistics, with the aim of offering students a foundation and framework for understanding more complex methods. Two application areas are considered throughout this text: (1) human genetic investigations in population-based association studies of unrelated individuals and (2) studies aiming to characterize associations between Human Immunodeficiency Virus (HIV) genotypes and phenotypes, as measured by *in vitro* drug responsiveness. Several publicly available datasets are used for illustration and can be downloaded at the book website (http://people.umass.edu/foulkes/asg.html). While data simulations are not described, emphasis is placed on understanding the implicit modeling assumption generally required for testing. An overarching theme of this text is that the application of any statistical method aims to characterize a *specific* relationship among variables. For example, just as an additive model of association can be used to evaluate additive structure, a classification or regression tree aims to characterize conditional associations. The array of methods that are applied to data arising from genetic association studies differ primarily in the types of associations that they are designed to uncover.

This text is also intended to complement the existing literature on statistical genetics and molecular epidemiology in two ways. First, this text offers extensive and integrated examples using R, an open-source, publicly available statistical computing software environment. This is intended both as a pedagogical tool for providing readers with a deeper understanding of the statistical algorithms presented and as a practical tool for applying the approaches described herein. Second, this text provides comprehensive coverage of both genetic concepts, such as linkage disequilibrium and Hardy-Weinberg equilibrium, from a statistical perspective, as well as fundamental statistical concepts, such as adjusting for multiplicity and methods for high-dimensional data analysis, relevant to the analysis of data arising from genetic association studies. Several excellent texts, including Thomas (2004) and Ziegler and Koenig (2007), provide in-depth coverage of genetic data concepts relevant to both population-based and family-based investigations. The present text presents these concepts within the context of familiar statistical nomenclature while providing coverage of several additional pertinent epidemiological concepts and statistical methods for characterizing association. This presentation is at a level that is accessible to the reader with a limited background in biostatistics and with an interest in public health or biomedical research. More advanced discussions of the underlying theory can be found in alternative texts such as Hastie *et al.* (2001) and Lange (2002), as well as the original manuscripts cited throughout this text.

The primary focus of this text is on candidate gene studies that involve the investigation of polymorphisms at several genetic sites within and across one or more genes and their associations with a trait. In the past several years, technological advancements leading to development and widespread availability of "SNP chips" have led to an explosion of genome-wide association studies (GWAS) involving 500 thousand to 1 million single-nucleotide polymorphisms (SNPs). The methods presented in this text apply equally to candidate gene approaches and whole and partial GWAS. Notably, however, the latter setting requires additional consideration of the computational burden of associated analysis as well as data preprocessing and error checking, as discussed in Section 3.3 and throughout this text. While GWAS have gained a great deal of popularity in recent years, they do not obviate the need for candidate gene studies that further investigate the role of specific genes in disease progression as well as the potential confounding or modifying roles of traditional risk factors, including both clinical and demographic characteristics. Instead, GWAS provide investigators with a vastly improved body of scientific knowledge to inform the selection of candidate genes for hypothesis-driven research.

The term high-dimensional has taken on many meanings across different fields of research and over the past decade of rapid expansion in these fields. In this text, high-dimensional is defined simply as a large number of potentially correlated variables that may interact, in a statistical or a biological sense, in their association with the outcome under investigation. The term is used loosely to refer to any number of variables for which there is a complex, uncharacterized structure and the usual least squares regression setting may not be easily applicable. High-dimensional data methods including approaches to multiplicity and characterizing gene–gene and gene–environment interactions are addressed within the context of characterizing associations among genetic sequence data and disease traits. In these settings, the predictor variables are SNPs or corresponding amino acids and are categorical. Primary consideration is given to dependent variables that are either continuous measures of disease progression or binary indicators of disease status, though brief mention is also made of methods for multivariate and survival outcomes. Specific attention is given to the potential confounding and mediating roles of individual-level clinical and demographic data.

Implementation of all described methods is demonstrated using the R environment and associated packages, which are publicly available at the Comprehensive R Archive Network (CRAN) website (http://cran.r-project.org/). The decision to use R in this text over alternative programming languages is multifaceted. First, as a publicly available package, R is freely accessible to all readers and, importantly, students will continue to have access to R at all future personal and professional venues. As an open-source language, R also provides students with the opportunity to view code used to generate functions, serving as a valuable pedagogical tool for more programmatically minded learners. Another key advantage of R is that investigators who develop new statistical methodology often provide an accompanying R package for implementation through the CRAN website, providing users with almost immediate access to implementation of the most recently developed approaches. Finally, with the availability of contributed packages, the choice of method to apply rests with the user rather than with what a core development team of the programming language chooses to release.

While strongly preferable for the reasons mentioned above, use of R in this text does have the drawback from a pedagogical perspective that both the versions and packages are updated frequently. That is, we see a clear trade-off between accessibility and stability. In the process of writing this text, several changes in the packages described herein occurred, resulting in inconsistent outputs. While these inconsistencies have been resolved as of the present date, several more are likely to arise over the next several years. The reader is encouraged to visit the textbook website for information on these changes. All of the programming scripts in this text were written and tested for R version 2.7.1. Ascii text files with complete R code used for the examples in this textbook can be found on the textbook website. The files can be downloaded, or read directly into R using the source() function. For example, to source the code from Example 1.1, we can write the following at the R prompt:

```
> source("http://people.umass.edu/foulkes/asg/examples/1.1.r")
```

Additionally specifying print.eval=T in this function call will print the corresponding output. While the programs presented within this text are comprehensive, the novice reader can begin with the appendix for a brief introduction to some fundamental concepts relevant to programming in R. Several, more comprehensive, introductions to R are available, and the reader is encouraged to reference these texts as well, including Gentleman (2008), Spector (2008) and Dalgaard (2002), for additional programming tools and background.

I am grateful for the advice and support I have received in writing this text from many colleagues, students, friends and family members. I would especially like to thank my students and postdoctoral fellows, M. Eliot, X. Li, Y. Liu, Dr. B.A. Nonyane and Dr. K. Au, who spent many hours checking for notational and programming consistency as well as sharing in helpful discussions. I am indebted to all of the students in the fall 2008 semester of public health 690T at the University of Massachusetts, Amherst for their helpful suggestions and for bearing with me in the first run of this text. I am grateful for having a long-term friend and colleague in Dr. R. Balasubramanian, whose support and encouragement were pivotal in my decision to write this text. I am also thankful for the many conversations with Dr. D. Cheng and her willingness to share her extensive knowledge in applied statistics. I am obliged to Dr. M.P. Reilly for an enduring collaboration that has fueled my interest and enhanced my knowledge in applied statistical genetics for medical research. I am grateful to Dr. A.V. Custer, whose dedication to the open-source software community was inspirational to me. Dr. V. De Gruttola's early mentorship continues to shape my research interests, and I am thankful for the passion and deep thinking he brings to our profession. I also value the strong encouragement and intellectual engagement of my early career mentors Dr. E. George and Dr. T. Ten Have. The efforts of Dr. E. Hoffman, Dr. H. Gorski and colleagues in providing the FAMuSS and HGDP data were extraordinary, and their commitment to public access to data resources is truly outstanding. I am also indebted to Dr. R. Shafer and colleagues for their remarkable effort in creating and maintaining the Stanford University HIV Drug Resistance Database, from which the Virco data were downloaded and several additional data sets can be accessed easily. I also greatly appreciate the insightful leadership of the R core development team and the individuals who wrote and maintain the R packages used throughout this text. All figures in this text were generated in R or created using the open-source graphics editor Inkscape (http://www.inkscape.org/). I value the many insightful comments and suggestions of the editors and anonymous reviewers. Support for this text was provided in part by a National Institute of Allergies and Infectious Disease (NIAID) individual research award (R01AI056983). Finally, thanks to my family for their tremendous love and support.

Andrea S. Foulkes
*Amherst, MA*
May 2009

# Contents

# List of Tables

# List of Figures

# Acronyms

AA: Amino acid

AIDS: Acquired immunodeficiency syndrome

ANOVA: Analysis of variance

BART: Bayesian additive regression tree

BSS: Between-group sum of squares

B-H: Benjamini and Hochberg (approach to multiple testing)

B-Y: Benjamini and Yekutieli (approach to multiple testing)

BMI: Body mass index

BVS: Bayesian variable selection

CART: Classification and regression trees

CV: Cross-validation

DNA: Deoxyribonucleic acid

EM: Expectation-maximization

FAMuSS: Functional SNPS Associated with Muscle Size and Strength

FDR: False discovery rate

FSDR: Free step-down resampling

FWEC: Family-wise error under the complete null

FWEP: Family-wise error under a partial null

FWER: Family-wise error rate

GLM: Generalized linear model

GWAS: Genome-wide association study

GWS: Genome-wide scan

HGDP: Human Genome Diversity Project

HTR: Haplotype trend regression

HWD: Hardy-Weinberg disequilibrium

HWE: Hardy-Weinberg equilibrium

IBD: Identical by descent

IBS: Identical by state

IDV: Indinavir

LD: Linkage disequilibrium

LOH: Loss of heterozygosity

LS: Learning sample

MARS: Multivariate adaptive regression splines

MCMC: Markov chain Monte Carlo

MDS: Multidimensional scaling

MI: Multiple imputation

MIRF: Multiple imputation and random forests

MSE: Mean square error

NFV: Nelfinavir

OOB: Out-of-bag

PCA: Principal components analysis

pFDR: Positive false-discovery rate

Pr: Protease

PRD: Positively regression dependent

QTL: Quantitative trait loci

RF: Random forest

RNA: Ribonucleic acid

RT: Reverse transcriptase

SAM: Significance analysis of microarrays

SNP: Single-nucleotide polymorphism

STP: Simultaneous test procedure

WSS: Within-group sum of squares

# 1

# Genetic Association Studies

Recent technological advancements allowing for large-scale sequencing efforts present an exciting opportunity to uncover the genetic underpinnings of complex diseases. In an attempt to characterize these genetic contributors to disease, investigators have embarked in multitude on what are commonly referred to as *population-based genetic association studies*. These studies generally aim to relate genetic sequence information derived from unrelated individuals to a measure of disease progression or disease status. The field of genomics spans a wide array of research areas that involve the many stages of processing from genetic sequence information to protein products and ultimately the expression of a trait. The breadth of genomic investigations also includes studies of multiple organisms, ranging from bacteria to viruses to parasites to humans. In this chapter, two settings are described in which population-based genetic association studies have marked potential for uncovering disease etiology while elucidating new approaches for targeted, individualized therapeutic interventions: (1) complex disease association studies in humans; and (2) studies involving the Human Immunodeficiency Virus (HIV).

In both settings, interest lies in characterizing associations between multiple genetic polymorphisms and a measured trait. In addition, these settings share the essential need to account appropriately for patient-level covariates as potential confounders or modifiers of disease progression to make clinically meaningful conclusions. While these two settings are not comprehensive, together they provide a launching point for discussion of quantitative methods that address the challenges inherent in many genetic investigations. This chapter begins by describing types of population-based studies, which represent one class of investigations within the larger field of genomics research. Also discussed are the fundamental features of data arising from these investigations as well as the analytical challenges inherent in this endeavor.

A.S. Foulkes, *Applied Statistical Genetics with R: For Population-based Association Studies*, Use R, DOI: 10.1007/978-0-387-89554-3_1,
© Springer Science+Business Media LLC 2009

## 1.1 Overview of population-based investigations

Population-based genetic association studies can be divided roughly into four categories of studies: candidate polymorphism, candidate gene, fine mapping and whole or partial genome-wide scans. In the following paragraphs, each of these types of studies is described briefly, followed by a discussion of how population-based genetic investigations fit within a larger context of genomic-based studies. Further discussions of population-based and family-based designs can be found in Thomas (2004) and Balding (2006).

### 1.1.1 Types of investigations

*Candidate polymorphism studies*

Investigations of genotype–trait associations for which there is an *a priori* hypothesis about functionality are called *candidate polymorphism* studies. Here the term *polymorphism* is defined simply as a genetic variant at a single location within a gene. Technically, a variation must be present in at least 1% of a population to be classified as a polymorphism. Such a variable site is commonly referred to as a single-nucleotide polymorphism (SNP). Candidate polymorphism studies typically rely on prior scientific evidence suggesting that the set of polymorphisms under investigation is relevant to the disease trait. The aim is to test for the presence of association, and the primary hypothesis is that the variable site under investigation is *functional*. That is, the goal of candidate polymorphism studies is to determine whether a given SNP or set of SNPs influences the disease trait directly.

*Candidate gene studies*

*Candidate gene* studies generally involve multiple SNPs within a single gene. The choice of SNPs depends on defined linkage disequilibrium (LD) blocks and is discussed further in Section 3.1. The underlying premise of these studies is that the SNPs under investigation capture information about the underlying genetic variability of the gene under consideration, though the SNPs may not serve as the true disease-causing variants. That is, the SNPs that are being studied are not necessarily functional. Consider for example a setting in which we want to investigate the association between a gene and disease. A gene comprises a region of deoxyribonucleic acid (DNA), representing a portion of the human genome. This is illustrated by the shaded rectangle in Figure 1.1. In a simple model, we might assume that a mutation at a single site within this region results in disease. In general, the precise location of this disease-causing variant is not known. Instead, investigators measure multiple SNPs that are presumed "close" to this site on the genome. The term "close" can be thought of as physical distance, though precise methods for choosing appropriate SNPs are described in more detail in Section 3.1.

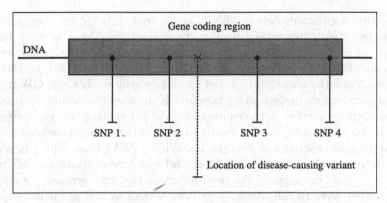

**Fig. 1.1.** Marker SNPs

These proximate SNPs are commonly referred to as *markers* since the observed genotype at these locations tends to be associated with the genotype at the true disease-causing locus. The idea underlying this phenomenon is that, over evolutionary time (that is, over many generations of reproduction), the disease allele was inherited alongside variants at these marker loci. This occurs when the probability of a recombination event in the DNA region between the disease locus and the marker locus is small. Thus, capturing variability in these loci will tend to capture variability in the true disease locus. Further discussion of recombination is provided in Section 1.3.1.

*Fine mapping studies*

The aim of *fine mapping* studies tends to differ from those of candidate gene and candidate polymorphism approaches. Fine mapping studies set out to identify, with a high level of precision, the *location* of a disease-causing variant. That is, these studies aim to determine precisely where on the genome the mutation that causes the disease is positioned. Knowledge about this location can obviate the need for investigations based on marker loci, thus reducing the error and variability in associated tests. Within the context of mapping studies, the term *quantitative trait loci (QTL)* is used to refer to a chromosomal position that underlies a trait. Methods for mapping and characterizing QTLs based on controlled experiments of inbred mouse lines are described in Chapter 15 of Lynch and Walsh (1998). Mapping studies are not a focal point of this text; however, we note that in some contexts the term "mapping" is used more loosely to refer to association, the topic of this text, in both family- and population-based studies. For comprehensive and advanced coverage of gene mapping methods, the reader is referred to Siegmund and Yakir (2007).

*Genome-wide association studies (GWAS)*

Similar to candidate gene approaches, studies involving whole and partial genome-wide scans, termed *genome-wide association studies (GWAS)*, aim to identify associations between SNPs and a trait. GWAS, however, tend to be less hypothesis driven and involve the characterization of a much larger number of SNPs. Partial scans generally involve between 100Kb and 500Kb segments of DNA, while whole-genome scans range from 500Kb to 1000Kb regions. While the underlying goal of candidate gene studies and GWAS can be similar, the data preprocessing is generally more extensive and the computational burden greater in the context of GWAS, requiring the application of software packages designed specifically to address the high-dimensional nature of the data, as described in Section 3.3. While GWAS have gained in popularity in recent years due to the advent and widespread availability of "SNP chips", they do not obviate the need for candidate gene studies. Candidate gene studies serve to validate findings from GWAS as well as further explore the biological and clinical interactions between genes and more traditional risk factors for complex diseases, such as age, gender, and other patient-level clinical and demographic characteristics. Importantly, the fundamental statistical concepts and methods described throughout this text are broadly relevant to both candidate gene studies and GWAS.

### 1.1.2 Genotype versus gene expression

The term "association" study has come to refer to studies that consider the relationship between genetic *sequence* information and a phenotype. Gene expression studies, based on microarray technology, on the other hand, aim to characterize associations among gene *products*, such as ribonucleic acid (RNA) or proteins, and disease outcomes. While the scientific findings from these investigations will likely lend support to one another, it is important to recognize that the two types of studies focus on different aspects of the cell life cycle. In the context of association studies, the raw genetic information as characterized by the DNA sequence is the primary predictor variable under investigation, and the aim is to understand how polymorphisms in the sequences explain the variability in a disease trait. Gene expression studies, on the other hand, focus on the extent to which a DNA sequence coding for a specific gene is transcribed into RNA (transcriptomics) and then translated into a protein product (proteomics). The former arises from gene chip technology and is commonly referred to as expression data, while the latter is an output of mass spectrometry. Since transcription and translation depend on many internal and external regulation factors, the expression of a gene sequence represents a different phenomenon than the sequence itself.

A fundamental unit of analysis in population association studies is the *genotype*. As described in Section 1.2, genotype is a categorical variable that takes on values from a predefined set of discrete characters. For example, in humans, most SNPs are *biallelic*, indicating there are two possible bases at the corresponding site within a gene (e.g., $A$ and $a$). Furthermore, since humans are *diploid*, each individual will carry two bases, corresponding to each of two homologous chromosomes. As a result, the possible genotype values in the population are $AA$, $Aa$ and $aa$. In studies of gene expression, on the other hand, the basic unit used in analysis is the gene product, which is typically a real-valued positive number. Notably, investigators may subsequently dichotomize this variable, though this additional level of data processing will depend on the scientific questions under consideration and prior knowledge.

In both settings, a measure of disease status or disease progress, referred to as the *trait* in this text, is also collected for analysis. Notably, in population association studies, we generally treat the genotype as the *predictor* variable and the trait as the *dependent* variable. In gene expression studies, this may or may not be the case. Consider for example the setting in which investigators aim to uncover the association between breast cancer and gene expression. In this case, the expression of a gene, as measured by how much RNA is produced, may serve as the main dependent variable, with cancer status as the potential predictor. The alternative formulation is also tenable. In this text, since emphasis is on population-based association studies, it is always assumed that genotype precedes the trait in the causal chain.

While careful consideration must be given to the several notable differences in the form as well as the interpretation of the data, many of the statistical methods described herein are equally applicable to gene expression studies. In the context of genotype data, we might for example test the null hypothesis that cholesterol level is the same for individuals with genotype $AA$ and genotype $aa$. In the expression setting, the null hypothesis may instead be framed as the gene expression level is the same for individuals with cardiovascular disease and those without cardiovascular disease. In both cases, a two-sample test for equality of means or medians (e.g., the two-sample $t$-test or Wilcoxon rank sum test) could be performed and similar approaches to account for multiple testing employed. Notably, preprocessing of gene expression data prior to formal statistical analysis also has its unique challenges. Several seminal texts provide discussion of statistical methods for the analysis of gene expression data. See for example Speed (2003), Parmigiani *et al.* (2003), McLachlan *et al.* (2004), Gentleman *et al.* (2005) and Ewens and Grant (2006).

Finally, we distinguish between genetic association studies and the rapidly growing field of research in epigenetics. The term *epigenetics* is used to describe heritable features that control the functioning of genes within an individual cell but do not constitute a physical change in the corresponding DNA sequence. The *epigenome*, defined literally as "above-the-genome", also referred to as the *epigenetic code*, includes information on methylation and histone patterns, called *epigenetic tags*, and plays an essential role in controlling the expression of genes. These tags can inhibit and silence genes, leading to common complex diseases such as cancer. In this text, we consider traditional epidemiological risk factors, such as smoking status and diet, that may play a role in defining an individual's epigenetic makeup; however, we do not address directly the challenges of epigenetic data. For a further discussion of the role of epigenetics in the link between environmental exposures and disease phenotypes, see Jirtle and Skinner (2007).

### 1.1.3 Population-versus family-based investigations

The term "population"-based is used to refer to investigations involving unrelated individuals and distinguished from family-based studies. The latter,
as the name implies, involves data collected on multiple individuals within
the same family unit. The statistical considerations for family-based studies
differ from those of population-based investigations in two primary regards.
First, individuals within the same family are likely to be more similar to one
another than are individuals from different families. This phenomenon is referred to in statistics as clustering and implies a within-family correlation.
The idea is that there is something unmeasurable (latent), such as diet or
underlying biological makeup, that makes people from the same family more
alike than people across families. As a result, the trait under investigation is
more highly correlated among individuals within the same family. Accounting for the potential within-cluster correlation in the statistical analysis of
family-based data is essential to making valid inference in these settings.

In population-based studies, a fundamental assumption is that individuals are unrelated; however, other forms of clustering may exist. For example,
individuals may have been recruited across multiple hospitals so that patients
from the same hospital are more similar than those across hospitals. This
within-cluster correlation can arise particularly if the catchment areas for the
hospitals include different socioeconomic statuses or if the standards for patient care are remarkably different. Alternatively, we may have repeated measurements of a trait on the same individual. This is another common situation
in which the assumption of independence is violated. In all of these cases, analytical methods for correlated data are again warranted and are essential for
correctly estimating variance components. In this text, attention is restricted
primarily to methods for independent observations, though consideration is
given to clustered data methods in Section 4.4.2. Tests for relatedness are also
described in Section 3.3. In-depth and comprehensive coverage of correlated
data methods can be found in Diggle *et al.* (1994), Vonesh and Chinchilli
(1997), Verbeke and Molenberghs (2000), Pinheiro and Bates (2000), McCulloch and Searle (2001), Fitzmaurice *et al.* (2004) and Demidenko (2004).

A second remarkable difference between population- and family-based
studies involves what is termed *allelic phase* and is defined as the alignment
of nucleotides on a single homolog. Allelic phase is typically unobservable in
population-based association studies but can often be determined in the context of family studies. This concept is described in greater detail in Section 1.2
and Chapter 5. As a result of these differences in the data structure, the methods for analysis of family-based association studies tend to differ from those
developed in the context of population-based studies. Though some of the
methods described herein, particularly adjustments for multiplicity, are applicable to family-based studies, this text focuses on methods specifically relevant
to population association studies, including inferring haplotypic phase (Chapter 5). Elaboration on the specific statistical considerations and methods for

family-based studies can be found in Khoury *et al.* (1993), Liu (1998), Lynch and Walsh (1998), Thomas (2004), Siegmund and Yakir (2007) and Ziegler and Koenig (2007).

### 1.1.4 Association versus population genetics

Finally, we distinguish between population-based association studies (the topic of this text) and *population genetic* investigations. Population genetics refers generally to the study of changes in the genetic composition of a population that occur over time and under evolutionary pressures. This includes, for example, the study of natural selection and genetic drift. In this text, we instead focus on estimation and inference regarding the association between genetic polymorphisms and a trait. Statistical methods relevant to population genetics are described in a number of texts, including Weir (1996), Gillespie (1998) and Ewens and Grant (2006).

## 1.2 Data components and terminology

Data arising from population-based genetic association studies are generally comprised of three components: (1) the *genotype* of the organism under investigation; (2) a single trait or multiple *traits* (also referred to as *phenotypes*) that are associated with disease progression or disease status; and (3) patient-specific *covariates*, including treatment history and additional clinical and demographic information. The primary aim of many association studies is to characterize the relationship between the first two of these components, the genotype and a trait. *Pharmacogenomic* investigations aim specifically to analyze how genotypes modify the effects of drug exposure (the third data component) on a trait. That is, these investigations focus on the statistical interaction between treatment and genotype on a disease outcome. While the specific aims of many association studies do not expressly involve the third data component, patient-specific clinical and demographic information, careful consideration of how these factors influence the relationship between the genotype and trait is essential to making valid biological and clinical conclusions. In this chapter, we describe each of these data components, all of which are highly relevant to population-based association studies, and introduce some additional terminology. A discussion of the potential interplay among components of the data and important epidemiological principles, including confounding, effect mediation, effect modification and conditional association, is provided in Section 2.1.2. Further elaboration on the concept of phase ambiguity and appropriate statistical approaches to handling this aspect of the data are given in Chapter 5.

### 1.2.1 Genetic information

Throughout this text, the term *genotype* is defined as the observed genetic sequence information and can be thought of as a categorical variable. The term *observed* is used here to distinguish genotype information from haplotype data, as described below. Humans carry two *homologous chromosomes*, which are defined as segments of deoxyribonucleic acid (DNA), one inherited from each parent, that code for the same trait but may carry different genetic information. Thus, in its rawest form in humans, the genotype is the pair of DNA bases adenine (A), thymine (T), guanine (G) and/or cytosine (C) observed at a location on the organism's genome. This pair includes one base inherited from each of the two parental genomes and should not be confused with the pairing that occurs to form the DNA double helix. These two types of pairing are described further in Section 1.3.1. Genotype data can take different forms across the array of genetic association studies and depend both on the specific organism under investigation and the scientific questions being considered, as we will see throughout this text.

The term *nucleotide* refers to a single DNA base linked with both a sugar molecule and phosphate and is often used interchangeably with the term DNA *base*. *Genes* are defined simply as regions of DNA that are eventually made into proteins or are involved in the regulation of transcription; that is, regions that regulate the production of proteins from other segments of DNA. In *candidate gene* studies, the set of genes under investigation is chosen based on known biological function. These genes may, for example, be involved in the production of proteins that are important components of one or more pathways to disease. In whole and partial *genome-wide association studies (GWAS)*, segments of DNA across large regions of the genome are considered and may not be accompanied by an *a priori* hypothesis about the specific pathways to disease.

In population-based association studies, the fundamental unit of analysis is the single-nucleotide polymorphism (SNP). A *SNP* simply describes a single base pair change that is variable across the general population at a frequency of at least 1%. The term can also be used more loosely to describe the specific location of this variability. The overriding premise of association studies is that there exists variability in DNA sequences across individuals that captures information on a disease trait. Regions of DNA within and across genes are said to have *genetic variability* if the alleles within the region vary across a population. *Conserved* regions, on the other hand, exhibit no variability in a population. Take the simple example of a single base pair location within a gene. If the genotype at this site is *AA* for all individuals within the population, then this site is referred to as conserved. On the other hand, if *AA*, *Aa* and *aa* are observed, then this site is called variable. Here the letters *A* and *a* are used to represent the observed nucleotides (A, C, T or G). For example, *A* may represent adenine (A) and *a* may represent thymine (T). Further discussion of notation is provided in Section 2.1.1. Highly conserved regions of DNA are less relevant in the context of association studies since they will not be able to capture the variability in the disease trait. Studying highly conserved regions would be tantamount in a traditional epidemiological investigation to only recruiting smokers to a study and then trying to assess the impact of smoking on cancer risk. Clearly, multiple levels of the predictor variable, in this case smoking status, are necessary if the goal is to assess the impact of this factor on disease.

*Multilocus genotype* is used to describe the observed genotype across multiple SNPs or genes, though the terms *genotype* and *multilocus genotype* are often used interchangeably. A *locus* or *site* can refer to the portion of the genome that encodes a single gene or the location of a single nucleotide on the genome. Multilocus genotype data consist of a string of categorical variables, with elements corresponding to the genotype at each of multiple sites on the genome. For example, an individual's multilocus genotype may be given by $(Aa, Bb)$, where $Aa$ is the genotype at one site and $Bb$ is the genotype at a second site. Again the letters $A$, $a$, $B$ and $b$ each represent the observed nucleotides (A, C, T or G). Notably, the specific ordering of alleles is non-informative, so, for example, the genotypes $Aa$ and $aA$ are equivalent.

The term multilocus genotype should not be confused with the concept of *haplotype*. Haplotype refers to the specific combination of alleles that are in *alignment* on a single *homolog*, defined as one of the two homologous chromosomes in humans. Suppose again that an individual's multilocus genotype is given by $(Aa, Bb)$. The corresponding pair of haplotypes, also referred to as this individual's *diplotype*, could be $(AB, ab)$ or $(Ab, aB)$. That is, either the $A$ and $B$ alleles are in alignment on the same homolog, in which case $a$ and $b$ align, or the $A$ and $b$ alleles align, in which case $a$ and $B$ are in alignment. These two possibilities are illustrated in Figure 1.2 and described further in Section 2.3.2. This uncertainty is commonly referred to as *ambiguity in allelic phase* or more simply *phase ambiguity*. In general, a multilocus genotype is observable, although missing data can arise from a variety of mechanisms. Haplotype data, on the other hand, are generally *unobservable* in population-based studies of unrelated individuals and require special consideration for analysis, as described in detail in Chapter 5.

This layer of missingness renders population-based association studies unique from family-based investigations. If parental information were available on the individual above, then it might be possible to clarify the uncertainty in allelic phase. For example, if the maternal genotype is $(AA, BB)$ and the paternal genotype is $(aa, bb)$, then it is clear that $A$ and $B$ align on the same homolog that was inherited from the maternal side and the $a$ and $b$ align on the copy inherited from the paternal side. In population-based studies, family data are generally not available to infer these haplotypes. However, it is possible to draw strength from the population haplotype frequencies to determine the most likely alignment for an individual. This is discussed in greater detail in Chapter 5.

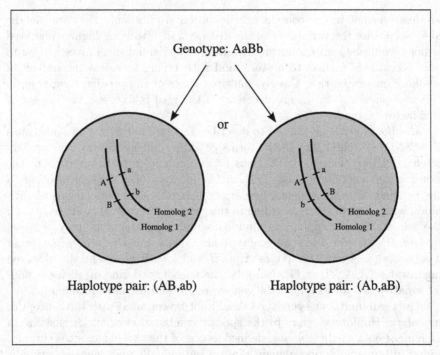

**Fig. 1.2.** Haplotype pairs corresponding to heterozygosity at two SNP loci

The term *zygosity* refers to the comparative genetic makeup of two homologous chromosomes. An individual is said to be *homozygous* at a given SNP locus if the two observed base pairs are the same. *Heterozygosity*, on the other hand, refers to the presence of more than one allele at a given site. For example, someone presenting with the *AA* or *aa* genotype would be called homozygous, while an individual with the *Aa* is said to be heterozygous at the corresponding locus. The term *loss of heterozygosity (LOH)*, commonly used in the context of oncology, refers specifically to the loss of function of an allele, when a second allele is already inactive, through inheritance of the heterozygous genotype.

The *minor allele* frequency, also referred to as the *variant allele* frequency, refers to the frequency of the less common allele at a variable site. Note that here the term *frequency* is used to refer to a population proportion, while statisticians tend to use the term to refer to a count. The terms *homozygous rare* and *homozygous variant* are commonly used to refer to homozygosity with two copies of the minor allele. Consider the simple example of a single-variable site for which *AA* is present in 75% of the population, *Aa* is present in 20% and *aa* is present in 5%. The frequency of the *A* allele is then equal to $(75 + 75 + 20)\%/2 = 85\%$, while the frequency of *a* is $(20 + 5 + 5)\%/2 = 15\%$. In this case, the minor allele (*a*) frequency is equal to 15%. The *major allele* is the more common allele and is given by $A$ in this example. An example of calculating the minor and major allele frequencies in R is provided in Section 1.3.3.

## 1.2.2 Traits

Population-based genetic association studies generally aim to relate genetic information to a *clinical outcome* or *phenotype*, which are both referred to in this text as a *trait*. The terms *quantitative* and *binary* traits refer respectively to continuous and binary variables, where a binary variable is defined as one that can take on two values, such as diseased or not diseased. The term *phenotype* is defined formally as a physical attribute or the manifestation of a trait and in the context of association studies generally refers to a measure of disease progression. In the context of viral genetic investigations, phenotypes typically refer to an *in vitro* measure such as the 50% inhibitory concentration ($IC_{50}$), which is defined as the amount of drug required to reduce the replication rate of the virus by 50%. The term *outcome* tends to mean the presence of disease, though it is often used more generally in a statistical sense to refer to any dependent variable in a modeling framework.

Clinical measures such as total cholesterol and triglyceride levels are examples of quantitative traits, while the indicator for a cardiovascular outcome, such as a heart attack, is an example of a binary trait. In a study of breast cancer, the trait may be defined as an indicator for whether or not a patient has breast cancer. In HIV investigations, traits include viral load (VL), defined as the concentration of virus in plasma, and CD4+ cell count, which is a marker for disease progression. In this text, the terms *trait, phenotype* and *outcome* are used broadly to refer to both *in vitro* and *in vivo* clinical measures of disease progression and disease status. Survival outcomes, such as the time to onset of AIDS, time to a cardiovascular event, or time to death, as well as ordinal outcomes, such as severity of disease, are other examples of traits that are also highly relevant to the study of genetic associations with disease. While this text focuses on continuous and binary traits, alternative formulations apply and the general methodology presented is applicable to a wider array of measures.

Traits can be measured cross-sectionally or over multiple time points spanning several weeks to several years. Data measured over time are referred to as *longitudinal* or *multivariate* data and provide several advantages from an analytical perspective, as discussed in detail in several texts, including Fitzmaurice *et al.* (2004). The choice of using cross-sectional or longitudinal data rests primarily on the scientific question at hand. For example, if interest lies in determining whether genotype affects the change in VL after exposure to a specific drug, then a longitudinal design with repeated measures of VL is essential. On the other hand, if the interest is in characterizing VL as a function of genotype at initiation of therapy, then cross-sectional data may be sufficient. While longitudinal studies can increase the power to detect association, they tend to be more costly than cross-sectional studies and are more susceptible to missing data and the resulting biases. In this text, we focus on the analysis of cross-sectional studies, though the overarching themes and concepts, such as multiple testing adjustments and the need to control type-1 error rates, are equally applicable to alternative modeling frameworks.

### 1.2.3 Covariates

In addition to capturing information on the genotype and trait, population-based studies generally involve the collection of other information on patient-specific characteristics. For example, in relating genetic polymorphisms to total cholesterol level among patients at risk for cardiovascular disease, additional relevant information may include body mass index (BMI), gender, age and smoking status. The additional data collected tend to be on variables that have previously been associated with the trait of interest, in this case cholesterol level, and may include environmental, demographic and clinical factors. Consideration of additional variables in the context of analysis will again depend on the scientific question at hand, the biological pathways to disease and the overarching goal of the analysis. For example, if the aim of a study is to identify the best predictive model (that is, to determine the model that can give the most accurate and precise prediction of cholesterol level for a new individual), then it is generally a good idea to include variables previously associated with the outcome in the model. If the goal is to characterize the association between a given gene and the outcome, then including additional variables, for example self-reported race, may also be warranted if these variables are associated with both the genotype and the outcome. This phenomenon is typically referred to as *confounding* and is discussed in greater detail in Chapter 2. On the other hand, if a variable such as smoking status is in the *causal pathway* to disease (that is, the gene under investigation influences the smoking status of an individual, which in turn tends to increase cholesterol levels), then inclusion of smoking status in the analysis may not be appropriate. In this text, the term *covariate* is used loosely to refer to any explanatory variables that are not of specific independent interest in the present investigation. Covariates are also commonly referred to as *independent* or *predictor* variables.

## 1.3 Data examples

Throughout this textbook, we provide examples using publicly available datasets, including data arising from two human-based investigations and one study involving HIV. Each of these datasets can be downloaded as ascii text files from the textbook website:

http://people.umass.edu/foulkes/asg.html

Below we include a summary of each dataset and example code for importing the data into R. Instructions for downloading R, inputing data and basic data manipulation strategies are given in the appendix. Additional elementary R concepts can be found in Gentleman (2008), Spector (2008), Venables and Smith (2008) and Dalgaard (2002). Complete information on all of the variables within each dataset can be found in the associated ReadMe files on the textbook website.

The two settings described in this section, complex disease association studies in humans and HIV genotype–trait association studies, serve as a framework for the methods presented throughout the text. While both the structure of the data and the overarching aims of the two settings are similar, there are a few notable differences worth mentioning. In both settings, belief lies in the idea that genetic polymorphisms (that is, variability in the genetic makeup across a population) will inform us about the variability observed in the occurrence or presentation of disease. Furthermore, this genetic variability in both HIV and humans is introduced through the process of replication. The *rate* at which these two organisms complete one life cycle, however, is dramatically different. While humans tend to replicate over the course of several years, an estimated $10^9$ to $10^{10}$ new virions are generated in a single day within an HIV-infected individual. Furthermore, the replication process for HIV, described in more detail below, is highly error-prone, resulting in a mutation rate of approximately $3 \times 10^{-5}$ per base per replication cycle, see for example Robertson *et al.* (1995).

As a result, there is a tremendous degree of HIV genetic variability within a single human host. That is, each HIV-infected individual carries an entire population of viruses, with each viral particle potentially comprised of different genetic material. In addition, the number of viral particles varies across individuals. Notably, both of these phenomena, genetic variability and the amount of virus in plasma, are influenced by current and past drug exposures. In contrast, humans carry two copies of each chromosome, with the exception of the sex chromosome, one inherited from each parent, and these *tend* to remain constant over an individual's lifetime. While relatively rare, mutations in the human genome do occur within a lifespan as a result of environmental exposure to mutagens. This process is notably slower in humans than in HIV and is not a focal point of this text. Additional details on each of these two settings are provided below.

## 1.3.1 Complex disease association studies

Characterizing the underpinnings of complex diseases, such as cardiovascular disease and cancer, is likely to require consideration of multiple genetic and environmental factors. As described in Section 1.1.1, human genetic investigations can involve several stages of processing of human genes, from the

DNA sequence to the protein product, and encompass a wide assortment of study designs. In this text, consideration is given to population-based studies of unrelated individuals, and the primary unit of genetic analysis is the DNA sequence. Humans inherit their genetic information from their two parental genomes through processes termed *mitosis* and *meiosis*. All human cells, with the exception of gametes, contain 46 chromosomes, including 22 homologous pairs, called *autosomes*, and 2 sex chromosomes. Each chromosome is comprised of a DNA double helix with two sugar-phosphate backbones connected by paired bases. In this context, guanine pairs with cytosine (G-C) and adenine pairs with thymine (A-T). This pairing is distinct from the pairing of homologous chromosomes that constitutes an individual's genotype. Notably, the latter pairing is not restricted, so that, for example, genotypes *GT* and *AC* can be observed.

*Mitosis* is a process of cell division that results in the creation of daughter cells that carry identical copies of this complete set of 46 chromosomes. *Meiosis* is the process by which a germ cell that contains 46 chromosomes, consisting of one homolog from each parent cell, undergoes two cell divisions, resulting in daughter cells, called *gametes*, with only 23 chromosomes each. In turn, this new generation of maternal and paternal gametes combines to form a *zygote*. A visual representation of meiosis is provided in Figure 1.3. Notably, prior to the meiotic divisions, each of the two homologous chromosomes are replicated to form *sister chromatid*. Subsequently, in the process of meiosis, cross-over between these maternal and paternal chromatids can occur. This is referred to as a *cross-over* or a *recombination event* and is depicted in the figure, where we see an exchange of segments of the paternal chromatid (shaded) and the maternal chromatid (unshaded). Finally, it is important to note that the 23 chromosomes are combined independently so that there are $2^{23} = 8,388,608$ possible combinations of chromosomes within a gamete. This phenomenon is commonly referred to as *independent assortment*. The reader is referred to any of a number of excellent textbooks that describe these processes in greater detail. See for example Chapter 19 of Vander *et al.* (1994) and Alberts *et al.* (1994).

Meiosis ensures two things: (1) each offspring carries the same number of chromosome pairs (23) as its parents; and (2) the genetic makeup of offspring is not identical to that of their parents. The latter results from both recombination and independent assortment. An important aspect of meiosis is that whole portions or *segments* of DNA within a chromosome tend to be passed from one generation to another. However, portions of DNA within chromosomes that are far from one another are less likely to be inherited together, as a result of recombination events. In the context of candidate gene studies, the SNPs under investigation can be known *functional* SNPs or what are referred to as *haplotype tagging* SNPs. Functional SNPs affect a trait directly, serving as a component within the causal pathway to disease. Haplotype tagging SNPs, on the other hand, are chosen based on their ability to capture overall variability within the gene under consideration. These SNPs tend to be associated

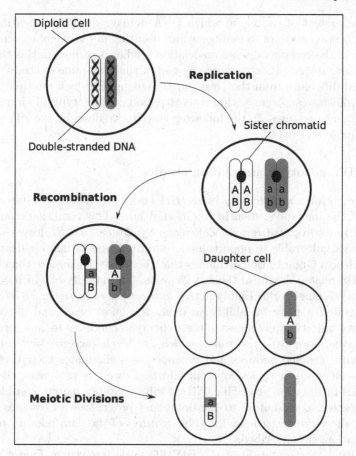

**Fig. 1.3.** Meiosis and recombination

with functional SNPs but may not be causal themselves. Notably, the length of a gene region can vary as well as the number of measured base pairs within each gene. The latter depends on what are called *linkage disequilibrium* blocks and relate to the probability of recombination within a region. This is described further in Section 3.1.

The structure of human genetics data is similar to that in the HIV setting, with a couple of notable exceptions. First, in human investigations, each individual has exactly two bases present at each location, one from each of the two homologous chromosomes. As described below in Section 1.3.2, in the viral genetics setting, an individual can be infected with multiple strains, resulting in any number of nucleotides at a given site. A second difference is that in many population-based association studies, human genetic sequence data are assumed to remain constant over the study period. One notable exception is in the context of cancer, in which DNA damage develops, resulting from environmental exposure to mutagens and resulting in uncontrolled cell proliferation. In the complex disease association studies described in this text, the genes under investigation do not vary within the timeframe of study. This is a marked difference from the viral genetic setting, in which multiple genetic polymorphisms can occur within a short period of time, typically in response to treatment pressures. In the following section, we describe the HIV genetic setting in greater detail.

### 1.3.2 HIV genotype association studies

The *Human Immunodeficiency Virus (HIV)* is a retrovirus that causes a weakening of the immune system in its infected host. This condition, commonly referred to as *Acquired Immunodeficiency Syndrome (AIDS)*, leaves infected individuals vulnerable to opportunistic infections and ultimately death. The World Health Organization estimates that there have been more than 25 million AIDS-related deaths in the last 25 years, the majority of which occurred in the developing world. Highly active *anti-retroviral therapies (ARTs)* have demonstrated a powerful ability to delay the onset of clinical disease and death, but unfortunately access to these therapies continues to be severely limited. Furthermore, drug resistance, which can be characterized by mutations in the viral genome, reduces and in some cases eliminates their usefulness. Both vaccine and drug development efforts, as well as treatment allocation strategies in the context of HIV/AIDS, will inevitably require consideration of the genetic contributors to the onset and progression of disease. In this section, the viral life cycle and notable features of the data relevant to these investigations are described.

A visual representation of the HIV life cycle is given in Figure 1.4. As a retrovirus, HIV is comprised of ribonucleic acid (RNA). From the figure, we see that the virus begins by fusing on the membrane of a CD4+ cell in the human host and injecting its core, which includes viral RNA, structural proteins, and enzymes, into the cell. The viral RNA is then reverse transcribed into DNA using one of these enzymes, *reverse transcriptase*. Another enzyme, *integrase*, then splices this viral DNA into the host cell DNA. The normal cell mechanisms for transcription and translation then result in the production of new viral protein. In turn, this protein is cleaved by the *protease* (Pr) enzyme and together with additional viral RNA forms a new virion. As this virion buds from the cell, the infected cell is killed, ultimately leading to the depletion of CD4 cells, which are vital to the human immune system. ARTs, the drugs used to treat HIV-infected individuals, aim to inhibit each of the enzymes involved in this life cycle.

Reverse transcription of RNA into DNA is a highly error-prone process, resulting in a mutation rate of approximately $3 \times 10^{-5}$ per base per cycle. This, coupled with a very fast replication cycle leading to $10^9$ to $10^{10}$ new virions each day, results in a very high level of genetic variability in the viral genome.

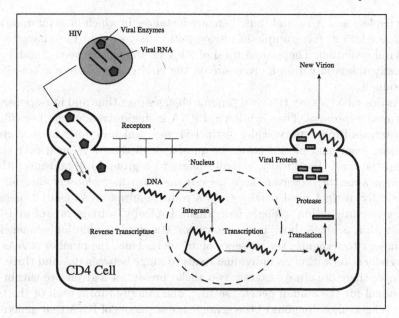

**Fig. 1.4.** HIV life cycle

The resulting viral population within a single human host is commonly referred to as a *quasi-species*. While many of these viruses are not viable (that is, they cannot survive with the resulting mutations), many others do remain. Notably, evidence suggests that mutated viruses can be transmitted from one host to another. The composition of a viral quasi-species tends to be highly influenced by current and past treatment exposures. HIV therapies generally consist of a combination of two or three anti-retroviral drugs, commonly referred to as a *drug cocktail*. There are currently four classes of drugs that each target a different aspect of the viral life cycle: fusion inhibitors, nucleoside reverse transcriptase inhibitors (NRTIs), non-nucleoside reverse transcriptase inhibitors (NNRTIs) and protease inhibitors (PIs). In the presence of these treatment pressures, viruses that are resistant to the drugs tend to emerge as the dominant species within a person. As individuals develop resistance to one therapy, another combination of drugs may be administered and a new dominant species can emerge. Evidence suggests that a blueprint of drug exposure history remains in latent reservoirs in the sense that a resistant species will re-emerge quickly in the presence of a drug to which a patient previously exhibited resistance.

The genetic composition of HIV is a single strand of RNA consisting of the four base pairs adenine (A), cytosine (C), guanine (G) and uracil (U). In general, and for the purpose of this textbook, the *amino acid* (AA) corresponding to three adjacent bases is of interest since AAs serve as the building blocks for proteins. Notably, there is not a one-to-one correspondence between base triplets and AAs, and thus there are instances in which base information is more relevant, for example in phylogenetic analyses aimed at characterizing viral evolution. There are a total of 20 AAs, though between 1 and 5 are typically observed within a given site on the viral genome across a sample of individuals.

As described above, the viral genome changes over time and in response to treatment exposures. Thus, while viral RNA is single stranded, an individual can carry multiple genotypically distinct viruses, which we refer to as *strains*, resulting from multiple infections or quasi-species that developed over time within the host. Technically, a strain refers to a group of organisms with a common ancestor; however, here we use the term more loosely to refer to genetically distinct viral particles. As a result, multiple AAs can be present at a given site within a single individual. Typically, a frequency of at least 20% within a single host is necessary for standard population sequencing technology to recognize the presence of an allele. Thus, the number of AAs at a given location within an individual tends to range between one and three. In contrast, there are always exactly two alleles present at a given site within an individual for the human genetic setting, one inherited from each of the two parental genomes. Regions of the genome are segments of RNA that generally code for a protein of interest. For example, in the context of studying viral resistance, the *Protease* (Pr) region and *Reverse Transcriptase* (RT) regions are of interest since these code for enzymes that are targeted by ARTs. The Envelope region, on the other hand, may be relevant to studies of vaccine efficacy since it is involved in cell entry. Regions are tantamount to genes in the context of human genetic studies.

### 1.3.3 Publicly available data used throughout the text

*The FAMuSS study*

The Functional SNPS Associated with Muscle Size and Strength (FAMuSS) study was conducted to identify the genetic determinants of skeletal muscle size and strength before and after exercise training. A total of $n = 1397$ college student volunteers participated in the study, and data on 225 SNPs across multiple genes were collected. The exercise training involved students training their non-dominant arms for 12 weeks. The primary aim of the study was to identify genes associated with muscle performance and specifically to understand associations among SNPs and normal variation in volumetric MRI (muscle, bone, subQ fat), muscle strength, response to training and clinical markers of metabolic syndrome. Primary findings are given in Thompson *et al.* (2004). A complete list of associated publications can be found in the ReadMe file on the textbook webpage.

The data are contained in a tab-delimited text file entitled `FMS_data.txt` and illustrated, in part, in Table 1.1. The file contains information on genotype across all SNPs as well as an extensive list of clinical and demographic factors

**Table 1.1.** Sample of FAMuSS data

| | fms.id | actn3_r577x | actn3_rs540874 | actn3_rs1815739 | actn3_1671064 | Term | Gender | Age | Race | NDRM.CH | DRM.CH |
|---|---|---|---|---|---|---|---|---|---|---|---|
| 1 | FA-1801 | CC | GG | CC | AA | 02-1 | Female | 27 | Caucasian | 40.00 | 40.00 |
| 2 | FA-1802 | CT | GA | TC | GA | 02-1 | Male | 36 | Caucasian | 25.00 | 0.00 |
| 3 | FA-1803 | CT | GA | TC | GA | 02-1 | Female | 24 | Caucasian | 40.00 | 0.00 |
| 4 | FA-1804 | CT | GA | TC | GA | 02-1 | Female | 40 | Caucasian | 125.00 | 0.00 |
| 5 | FA-1805 | CC | GG | CC | AA | 02-1 | Female | 32 | Caucasian | 40.00 | 20.00 |
| 6 | FA-1806 | CT | GA | TC | GA | 02-1 | Female | 24 | Hispanic | 75.00 | 0.00 |
| 7 | FA-1807 | TT | AA | TT | GG | 02-1 | Female | 30 | Caucasian | 100.00 | 0.00 |
| 8 | FA-1808 | CT | GA | TC | GA | 02-1 | | | | | |
| 9 | FA-1809 | CT | GA | TC | GA | 02-1 | Female | 28 | Caucasian | 57.10 | -14.30 |
| 10 | FA-1810 | CC | GG | CC | AA | 02-1 | Male | 27 | Hispanic | 33.30 | 0.00 |
| 11 | FA-1811 | CC | GG | CC | AA | 02-1 | | | | | |
| 12 | FA-1812 | CT | GA | TC | GA | 02-1 | Female | 30 | Caucasian | 20.00 | 0.00 |
| 13 | FA-1813 | CT | GA | TC | GA | 02-1 | Female | 20 | Caucasian | 25.00 | 25.00 |
| 14 | FA-1814 | CT | GA | TC | GA | 02-1 | Female | 23 | African Am | 100.00 | 25.00 |
| 15 | FA-1815 | | | | | | | | | | |
| 16 | FA-1816 | TT | GA | TC | GA | 02-1 | Female | 24 | Caucasian | 28.60 | 12.50 |
| 17 | FA-1817 | CT | GA | TC | GA | 02-1 | | | | | |
| 18 | FA-1818 | CT | GA | TC | GA | 02-1 | | | | | |
| 19 | FA-1819 | CT | GG | CC | AA | 02-3 | Male | 34 | Caucasian | 7.10 | -7.10 |
| 20 | FA-1820 | CC | GA | TC | GA | 02-3 | Female | 31 | Caucasian | 75.00 | 20.00 |

for a subset ($n = 1035$) of the study participants. We begin by specifying the web location of the data file as follows:

```
> fmsURL <- "http://people.umass.edu/foulkes/asg/data/FMS_data.txt"
```

We then use the `read.delim()` function to pull the data into R directly from the textbook website:

```
> fms <- read.delim(file=fmsURL, header=T, sep="\t")
```

By specifying `header=T`, we are indicating that the first row of the text file contains the variable names. Alternatively, we could have specified `header=F`, which assumes that the first line of the file is the first record of data. We also indicate with the argument `sep="\t"` that a tab separates each variable within a line of the data. Common alternative specifications are `sep=","` and `sep=""`, indicating comma and space delimiters, respectively. As described in the appendix, other useful functions for reading data into R include `read.table()` and `read.csv()`. The specifications given above are the default values for `read.delim()` and need not be written out explicitly. We do so for the purpose of illustration.

A portion of the data on the first 20 individuals in this sample are displayed in Table 1.1. Included in this table are the genotypes for four SNPs within the `actn3` gene and a few corresponding clinical and demographic parameters. The variable `Term` indicates the year and term (1—spring, 2—summer, 3—fall) of recruitment into the study, and `Gender`, `Age` and `Race` are all self-declared values of these demographic factors. The percentage changes in muscle strength before and after exercise training are given by `NDRM.CH` for the non-dominant arm and `DRM.CH` for the dominant arm. Generation of the LaTeX code for Table 1.1 is done in R using the `xtable()` function in the `xtable` package. The `print()` function with the `floating.environment` option set equal to `'sidewaystable'` is used to generate a landscape table. Alternatively, we can print the table in R as shown below:

```
> attach(fms)
> data.frame(id, actn3_r577x, actn3_rs540874, actn3_rs1815739,
+        actn3_1671064, Term, Gender, Age, Race, NDRM.CH,DRM.CH)[1:20,]
```

We use the `attach()` function so that we can call each variable by its name without having to indicate the corresponding dataframe. For example, after submitting the command `attach(fms)`, we can call the variable `Gender` without reference to `fms`. Alternatively, we could write `fms$Gender`, which is valid whether or not the `attach()` function was used. A dataframe must be re-attached at the start of a new R session for the corresponding variable names to be recognized. The numbers `1:20` within the square brackets and before the comma are used to indicate that row numbers 1 through 20 are to be printed.

We see from this table that the genotype for `id=FA-1801` at the first recorded SNP (`r577x`) within the gene `actn3` is the pair of bases $CC$. In most cases, SNPs are biallelic, which means that two bases are observed within a site across individuals. For example, for SNP r577x in gene actn3, the letters $C$ and $T$ are observed, while at rs540874 in gene actn3, the two bases $G$ and $A$ are observed. This pairing is not restricted (that is, $A$ can be present with $T$, $C$ or $G$ within another site), distinguishing this from the pairing of bases that occurs to form the DNA double helix within a single homolog (in which $A$ always pairs with $T$ and $C$ with $G$).

Recall that an individual is said to be *homozygous* if the two observed base pairs are the same at a given site and *heterozygous* if they differ. From Table 1.1, for example, we see that individual FA-1801 from the FAMuSS study is homozygous at actn3_rs540874 with the observed genotype equal to $GG$. Likewise, individual FA-1807 is homozygous at this site since the observed genotype is $AA$. Individuals FA-1802, 1803 and 1804, on the other hand, are all heterozygous at actn3_rs540874 since their genotypes contain both the $G$ and $A$ alleles. Determination of a minor allele and its frequency is demonstrated in the following example using data from the FAMuSS study.

*Example 1.1 (Identifying the minor allele and its frequency).* Suppose we are interested in determining the minor allele for the SNP labeled actn3_rs540874 in the FAMuSS data. To do this, we need to calculate corresponding allele frequencies. First we determine the number of observations with each genotype for this SNP using the following code:

```
> attach(fms)
> GenoCount <- summary(actn3_rs540874)
> GenoCount

  AA   GA   GG NA's
 226  595  395  181
```

The table() function in R outputs the counts of each level of the ordinal variable given as its argument. In this case, we see $n = 226$ individuals have the $AA$ genotype, $n = 595$ individuals have the $GA$ genotype and $n = 395$ individuals have the $GG$ genotype. An additional $n = 181$ individuals are missing this genotype. For simplicity, we assume that this missingness is non-informative. That is, we make the strong assumption that our estimates of the allele frequencies would be the same had we observed the genotypes for these individuals. To calculate the allele frequencies, we begin by determining our reduced sample size (that is, the number of individuals with complete data):

```
> NumbObs <- sum(!is.na(actn3_rs540874))
```

The genotype frequencies for $AA$, $GA$ and $GG$ are then given respectively by

```
> GenoFreq <- as.vector(GenoCount/NumbObs)
> GenoFreq

[1] 0.1858553 0.4893092 0.3248355 0.1488487
```

The frequencies of the $A$ and $G$ alleles are calculated as follows:

```
> FreqA <- (2*GenoFreq[1] + GenoFreq[2])/2
> FreqA
```

```
[1] 0.4305099
```

```
> FreqG <- (GenoFreq[2] + 2*GenoFreq[3])/2
> FreqG
```

```
[1] 0.5694901
```

Thus, we report $A$ is the minor allele at this SNP locus, with a frequency of 0.43. In this case, an individual is said to be homozygous rare at SNP rs540874 if the observed genotype is $AA$. *Homozygous wildtype*, on the other hand, refers to the state of having two copies of the more common allele, or the genotype $GG$ in this case.

Alternatively, we can achieve the same result using the genotype() and summary() functions within the genetics package. First we install and upload the R package as follows:

```
> install.packages("genetics")
> library(genetics)
```

We then create a genotype object and summarize the corresponding genotype and allele frequencies:

```
> Geno <- genotype(actn3_rs540874,sep="")
> summary(Geno)
```

```
Number of samples typed: 1216 (87%)
```

```
Allele Frequency: (2 alleles)
    Count Proportion
G   1385      0.57
A   1047      0.43
NA   362       NA
```

```
Genotype Frequency:
    Count Proportion
G/G   395      0.32
G/A   595      0.49
A/A   226      0.19
NA    181       NA
```

```
Heterozygosity (Hu)  = 0.4905439
Poly. Inf. Content   = 0.3701245
```

Here we again see that $A$ corresponds to the minor allele at this SNP locus, with a frequency of 0.43, while $G$ is the major allele, with a greater frequency of 0.57.                                                    □

*The Human Genome Diversity Project (HGDP)*

The Human Genome Diversity Project (HGDP) began in 1991 with the aim of documenting and characterizing the genetic variation in humans worldwide (Cann *et al.*, 2002). Genetic and demographic data are recorded on $n = 1064$ individuals across 27 countries. In this text, we consider genotype information across four SNPs from the v-akt murine thymoma viral oncogene homolog 1 (AKT1) gene. In addition to genotype information, each individual's country of origin, gender and ethnicity are recorded. For complete information on this study, readers are referred to http://www.stanford.edu/group/morrinst/ hgdp.html. Data are contained in the tab-delimited text file `HGDP_AKT1.txt` on the textbook website. Again we begin by specifying the location of the data:

```
> hgdpURL <- "http://people.umass.edu/foulkes/asg/data/HGDP_AKT1.txt"
```

Then we apply the `read.delim()` function to read the data into R:

```
> hgdp <- read.delim(file=hgdpURL, header=T, sep="\t")
```

Data on the first 20 observations in this dataset are provided in Table 1.2. Here the variable `Population` refers to ethnicity, `Geographic.origin` is the country of origin and `Geographic.area` is a more general description of location for the individuals in this cohort.

*The Virco data*

Several publicly available datasets that include viral sequence information, treatment histories and clinical measures of disease progression for HIV-infected individuals are downloadable at the Stanford Resistance Database: http://hivdb.stanford.edu/. In this text we consider a data set generated by $\text{Virco}^{TM}$, which includes protease (Pr) sequence information on 1066 viral isolates and corresponding fold-resistance measures for each of eight Pr inhibitors. Fold resistance is a comparative measure of responsiveness to a drug, where the referent value is for a *wildtype* or *consensus* virus. The consensus AA at a site on the viral genome is defined as the AA that is most common at this site in the general population. The data are comma delimited and contained in the file `Virco_data.csv` on the textbook website. We use the `read.csv()` function in R to read in the data:

```
> vircoURL <- "http://people.umass.edu/foulkes/asg/data/Virco_data.csv"
> virco <- read.csv(file=vircoURL, header=T, sep=",")
```

Note that we now indicate `sep=","` since the data are comma delimited. This is the default for the `read.csv()` function. Complete information on the variables in the database and associated publications can be found on the Stanford Resistance Database website. A sample of the data on a select set

Table 1.2. Sample of HGDP data.

| | Well ID | Gender | Population | Geographic.origin | Geographic.area | AKT1 C0756A | C6024T | G2347T | G2375A |
|---|---|---|---|---|---|---|---|---|---|
| 1 B12 | HGDP00980 | F | Biaka Pygmies | Central African Republic | Central Africa | CA | CT | TT | AA |
| 2 A12 | HGDP01406 | M | Bantu | Kenya | Central Africa | CA | CT | TT | AA |
| 3 E5 | HGDP01266 | M | Mozabite | Algeria (Mzab) | Northern Africa | AA | TT | TT | AA |
| 4 B9 | HGDP01006 | F | Karitiana | Brazil | South America | AA | TT | TT | AA |
| 5 E1 | HGDP01220 | M | Daur | China | China | AA | TT | TT | AA |
| 6 H2 | HGDP01288 | M | Han | China | China | AA | TT | TT | AA |
| 7 G3 | HGDP01246 | M | Xibo | China | China | AA | TT | TT | AA |
| 8 H10 | HGDP00705 | M | Colombian | Colombia | South America | AA | TT | TT | AA |
| 9 H11 | HGDP00706 | F | Colombian | Colombia | South America | AA | TT | TT | AA |
| 10 H12 | HGDP00707 | F | Colombian | Colombia | South America | AA | TT | TT | AA |
| 11 A2 | HGDP00708 | F | Colombian | Colombia | South America | AA | TT | TT | AA |
| 12 A3 | HGDP00709 | M | Colombian | Colombia | South America | AA | TT | TT | AA |
| 13 A4 | HGDP00710 | M | Colombian | Colombia | South America | AA | TT | TT | AA |
| 14 F5 | HGDP00598 | M | Druze | Israel (Carmel) | Israel | AA | TT | TT | AA |
| 15 G11 | HGDP00684 | F | Palestinian | Israel (Central) | Israel | AA | TT | TT | AA |
| 16 C2 | HGDP00667 | F | Sardinian | Italy | Southern Europe | AA | TT | TT | AA |
| 17 E10 | HGDP01155 | M | North Italian | Italy (Bergamo) | Southern Europe | AA | TT | TT | AA |
| 18 B7 | HGDP01415 | M | Bantu | Kenya | Central Africa | AA | TT | TT | AA |
| 19 B8 | HGDP01416 | M | Bantu | Kenya | Central Africa | AA | TT | TT | AA |
| 20 G4 | HGDP00865 | F | Maya | Mexico | Central America | AA | TT | TT | AA |

of variables is given in Table 1.3. The variable `SeqID` is the sequence iden-
tifier, and `IsolateName` is the name given to the corresponding isolate. The
drug-specific fold-resistance variables are labeled `Drug.Fold`, so, for example,
Indinavir (IDV) fold resistance is given by the variable `IDV.Fold`. A higher
fold-resistance value indicates that the corresponding isolate is more resistant
(less sensitive) to the indicated drug than a wildtype sequence based on an *in
vitro* assay.

The genotype information is available in two formats. The first representation is given by the variables with names that begin with the letter `P` and
followed by a number. This number refers to the amino acid position within
the Pr region of the viral sequence. For example, the variable `P10` represents
the tenth AA position within the Pr region of the viral genome. A "−" in
the data table indicates the presence of the population consensus AA, while a
letter indicates a mutation in the form of the AA corresponding to this letter.
For example, for `SeqID==3852`, a variant AA is observed at site 10 in the form
of Isoleucine ($I$). A total of 99 P variables are included in this dataset, corresponding to the 99 AA sites in the protease region of the viral genome. An
alternative formulation of the data is given by the variable `MutList`, which is
a list of all the observed mutations. These data are coded by a letter, followed
by a number, followed by another letter. The number is again the AA location, the first letter is the consensus AA at this site and the letter following
the number is the AA(s) that are observed at the corresponding location. For
example, $L10I$ indicates that AA $I$ is present in place of leucine ($L$) at site
10.

**Table 1.3.** Sample Virco data

| | SeqID | IsolateName | IDV.Fold | P10 | P63 | P71 | P82 | P90 | CompMutList |
|---|---|---|---|---|---|---|---|---|---|
| 1 | 3852 | CA3176 | 14.20 | I | P | - | - | M | L10I, M46I, L63P, G73CS, V77I, L90M, I93L |
| 2 | 3865 | CA3191 | 13.50 | I | P | V | T | M | L10I, R41K, K45R, M46I, L63P, A71V, G73S, V77I, V82T, I85V, L90M, I93L |
| 3 | 7430 | CA9998 | 16.70 | I | P | V | A | M | L10I, I15V, K20M, E35D, M36I, I54V, R57K, I62V, L63P, A71V, G73S, V82A, L90M |
| 4 | 7459 | Hertogs-Pt1 | 3.00 | I | P | T | - | M | L10I, L19Q, E35D, G48V, L63P, H69Y, A71T, L90M, I93L |
| 5 | 7460 | Hertogs-Pt2 | 7.00 | - | - | - | A | - | K14R, I15V, V32I, M36I, M46I, V82A |
| 6 | 7461 | Hertogs-Pt3 | 21.00 | I | P | V | A | M | L10I, K20R, M36I, N37D, I54V, R57K, D60E, L63P, A71V, I72V, V82A, L90M, I93L |
| 7 | 7462 | Hertogs-Pt4 | 8.00 | - | P | - | A | - | M36I, G48V, I54V, D60E, I62V, L63P, V82A |
| 8 | 7463 | Hertogs-Pt5 | 100.00 | I | - | V | A | M | L10I, I13V, M36I, N37D, G48V, I54V, D60E, Q61E, I62V, I64V, A71V, V82A, L90M, I93L |
| 9 | 7464 | Hertogs-Pt6 | 18.00 | - | P | - | A | - | V32I, M46I, L63P, V82A, I93L |
| 10 | 7465 | Hertogs-Pt7 | 15.00 | - | I | V | A | M | E34K, R41K, K43R, I54V, I62V, L63I, A71V, T74S, V82A, L90M |
| 11 | 7466 | Hertogs-Pt8 | 4.00 | I | P | - | - | - | L10I, E35D, M36I, G48V, D60E, L63P, H69Y |
| 12 | 7467 | Hertogs-Pt9 | 45.00 | - | P | V | - | - | I13V, K14R, K20M, E35D, M36I, N37D, K45R, L63P, H69X, A71V, I84V, L89X |
| 13 | 15492 | RC-V33778 | 1.00 | X | - | V | - | - | L10X, I15V, I50V, I62V, A71V, I72V, N83Z |
| 14 | 15493 | RC-V213888 | 1.00 | F | A | - | - | - | L10F, I15V, L33F, M46X, I50V, L63A, T74S, V77I, L89M |
| 15 | 15494 | RC-V207648 | 2.00 | F | - | - | - | - | L10F, V32I, M46I, I47V, I62V |
| 16 | 15495 | RC-V022292 | 3.00 | - | P | V | A | M | E34Z, R41K, I54V, I62V, L63P, A71V, V82A, L90M, I93L |
| 17 | 15498 | RC-V020855 | 1.00 | I | X | X | - | - | L10I, G48V, I54X, L63X, I64V, A71X, I93L |
| 18 | 15499 | RC-V216965 | 1.00 | - | T | V | M | X | L33X, K43Z, M46V, I50V, Q58E, D60E, L63T, I64V, A71V, I72Z, Y77I, V82M, L90X |
| 19 | 15500 | RC-V020829 | 0.50 | I | P | - | - | - | L10I, D30N, E35D, M36V, P39Z, L63P, N88D, I93L |
| 20 | 15501 | RC-V020834 | 1.00 | - | P | - | - | M | E35D, M36I, G48V, L63P, H69Z, L90M |

# Problems

**1.1.** State the primary analytic considerations that distinguish population-based and family-based investigations.

**1.2.** Define and contrast the following terms: (a) genotype, (b) haplotype, (c) phase, (d) homologous, (e) allele, and (f) zygosity.

**1.3.** Based on the FAMuSS data, determine the minor allele and its frequency for the `actn3_1671064` SNP. Report these frequencies overall and stratified by the variable labeled `Race`. Interpret your findings.

**1.4.** Using the HGDP data, summarize the genotype frequencies for the SNP labeled `AKT1.C6024T`, overall and by geographic area, using the variable named `geographic.area`. Interpret the results.

**1.5.** Report the observed proportion of mutations at sites 1, 10, 30, 71, 82 and 90 in the Protease region of the HIV genome for the Virco data using the variables labeled P1, P10, P30, P71, P82 and P90. Explain your findings.

# 2

# Elementary Statistical Principles

This chapter includes coverage of several statistical and epidemiological concepts that are broadly relevant to the study of association among multiple variables and specifically important in the analysis of genetic association in population-based investigations. The chapter is divided into three sections. Section 2.1 offers a general background, including the notation used throughout this text and some elementary probability concepts. This section also provides a basic overview of several fundamental epidemiological concepts relevant to any population-based investigation, including confounding, effect mediation, effect modification and conditional association. The reader is referred to Rothman and Greenland (1998) for a comprehensive overview of these and other epidemiological principles. Some of these concepts are very similar to the genetic data concepts discussed in Chapter 3, though they tend to have different nomenclature. All of these elements are important to the discovery and characterization of genotype–trait associations. Section 2.2 describes several simple measures and tests of statistical association, including correlation analysis, contingency table analysis and simple linear and logistic regression. Also included in this section is an introduction to methods for multivariable analysis. Finally, Section 2.3 offers an overview of the analytic challenges inherent in population-based genetic investigations. The testing procedures described throughout this chapter can be applied to each of a set of genotype variables though they generally require an adjustment for multiple comparisons, as described in Chapter 4. Further extensions that allow simultaneous assessment of associations for a group of SNPs or genes are presented in Chapters 5–7.

A.S. Foulkes, *Applied Statistical Genetics with R: For Population-based Association*
*Studies*, Use R, DOI: 10.1007/978-0-387-89554-3_2,
© Springer Science+Business Media LLC 2009

## 2.1 Background

### 2.1.1 Notation and basic probability concepts

*Notation*

As described in Section 1.2, data arising from population-based association studies are comprised of three primary components: the trait, which in this text is either quantitative or binary; a single or multilocus genotype; and several patient-level covariates. Throughout this text, we use $y$ to represent the trait under study, $x$ to represent the genotype data and $z$ to represent covariates. For example, we let $y_i$ be the trait for the $i$th individual in our sample, where $i = 1, \ldots, n$ and $n$ is our total sample size. Furthermore, $x_{ij}$ is the genotype at the $j$th SNP for individual $i$, where $j = 1, \ldots, p$ and $p$ is the total number of SNPs under study. Finally, $z_{ik}$ is the value of the $k$th covariate for individual $i$, where $k = 1, \ldots, m$ and $m$ is the number of covariates.

Boldface font is used to indicate vectors, and capital letters are used to represent matrices. For example, we use $\mathbf{x} = (x_1, \ldots, x_n)^T$ to represent an $n \times 1$ vector of genotypes at a single site on the genome across all individuals in our sample and $\mathbf{x}_j = (x_{1j}, \ldots, x_{nj})^T$ to represent the genotypes at the $j$th site under study. The notation $T$ is used to indicate the transpose of a vector. Similarly, $\mathbf{y} = (y_1, \ldots, y_n)^T$ is a vector with its $i$th element corresponding to the trait for individual $i$. In several settings, we additionally use $\mathbf{x}_i = (x_{i1}, \ldots, x_{ip})^T$ to denote the genotype data for the $i$th individual across $p$ variables. The difference between $\mathbf{x}_i$ and $\mathbf{x}_j$ will be made clear by the particular context presented. Recall that $\mathbf{y}$ can be quantitative, such as CD4 count or total cholesterol, in which case it is typically measured with error. Alternatively, $\mathbf{y}$ may be a binary random variable representing disease status.

An $n \times p$ matrix of genotype variables is given by $\mathbf{X}$, with the $(i,j)$-element corresponding to the $j$th genotype for individual $i$. Similarly, the $n \times m$ matrix $\mathbf{Z}$ is used to represent the entire set of covariates. These may include multiple clinical, demographic and environmental variables, such as age, sex, weight and second-hand smoke exposures. The concatenated matrix given by $\begin{bmatrix} \mathbf{X} & \mathbf{Z} \end{bmatrix}$ represents all potential explanatory variables. If dimensions are not indicated, they can generally be inferred based on the specific model of association under consideration. Finally, while Roman letters are used to represent data, Greek symbols, such as $\alpha$, $\mu$, $\beta$ and $\theta$, are used to represent model parameters. These parameters are unobservable quantities that we are generally interested in estimating or making inference about.

Application of any statistical approach requires first understanding the potential ways in which the components of our data can be defined. In general, the genotype for individual $i$ at site $j$, denoted $x_{ij}$, is a categorical variable that takes on two or more levels. For example, $x_{ij}$ may be defined as a three-level factor variable taking on the three possible genotypes at a biallelic site, given by $AA$, $Aa$ and $aa$. Alternatively, we can define $x_{ij}$ as a binary variable indicating the presence of at least one variant allele at a single SNP locus. That is, for example, we can let $x_{ij} = 0$ if the observed genotype is $AA$ and $x_{ij} = 1$ otherwise. Yet another alternative is to define $x_{ij}$ as an indicator for the presence of any variant alleles across multiple SNPs within a given gene. For example, suppose there are two SNPs within the $j$th gene. We can then let $x_{ij} = 0$ if the multilocus genotype is $(AA, BB)$ and $x_{ij} = 1$ otherwise.

Note that throughout this text we offer simple examples in which the letters $A$, $a$, $B$ and $b$ are used to represent the alleles at two biallelic loci. The two letters $A$ and $B$ are used to indicate different sites, and the capital and lowercase letters are intended to represent the different alleles at each corresponding site. For example, if the nucleotides $C$ and $G$ are observed at the first site, then we may let $A = C$ and $a = G$. In general, $A$ represents the major allele and $a$ represents the minor allele. This notation is presented in several genetics texts, though alternative notation has also been used. Most notably, we often see the notation $A_1$, $A_2$, $B_1$ and $B_2$, where now the subscript is used to indicate the corresponding allele. The advantage of this alternative notation is that we are not restricted to biallelic SNPs. That is, if more than two alleles are present in a population at a given site, we can represent these as $A_1$, $A_2$, $A_3$, ..., $A_k$ for $k > 2$. In this text, we resort to the more commonly used $A$, $a$ notation since it tends to be less cumbersome within larger formulas.

*Statistical independence*

The concept of independence is a cornerstone of statistical inference. In the context of genetic association studies, we are often interested in testing the null hypothesis that the trait under study is independent of genotype. That is, a summary value of the trait, such as the mean, is the same for all levels of genotype. As we will see in Chapter 3, assessing independence of alleles across two or more loci (linkage equilibrium) or independence of alleles across two homologous chromosomes (Hardy-Weinberg equilibrium) is also of general interest in genetic association studies. In probability terms, we say that two events are *independent* if the joint probability of their occurrence is the product of the probabilities of each event occurring on its own.

Consider for example a single SNP with alleles given by $A$ and $a$. Suppose one event is defined as the presence of $A$ on one of the two homologous chromosomes and a second event is defined as the presence of $A$ on the second of the two homologous chromosomes. Further, let $p_A$ be the allele frequency of $A$ and $p_{AA}$ be the joint probability of having two copies of $A$, one on each of the two homologous chromosomes. We say these two events are *independent* (that is, the occurrences of $A$ on each of the two homologous chromosomes are independent) if $p_{AA} = p_A p_A$. For example, if the probability of $A$ is $p_A = 0.50$, then, under independence, the probability of the $AA$ genotype is $p_{AA} = 0.5^2 = 0.25$.

Dependency, on the other hand, refers to the situation in which the probability of one event depends on the outcome of another event. For example, suppose one event is defined as having a mutation in the BRCA1 gene, a known risk factor for breast and ovarian cancers, and a second event is defined as developing breast cancer. In this case, the probability of developing breast cancer depends on whether a mutation is present; that is, the second event depends on whether the first event occurred. Formally, let our two events be denoted $E_1$ and $E_2$. The conditional probability of the second event given that the first one has occurred is written $Pr(E_2|E_1)$, where $|$ is read "given". In general, we have $Pr(E_2|E_1) = Pr(E_2 \text{ and } E_1)/Pr(E_1)$. If the two events are independent, then this reduces to $Pr(E_2|E_1) = Pr(E_2)$, and similarly we have $Pr(E_1|E_2) = Pr(E_1)$.

*Expectation*

The expectation of a discrete random variable can be thought of simply as a weighted average of its possible values, where the weights are equal to the probabilities that the variable takes on corresponding values. For example, suppose $Y$ is a random variable that takes on the value 1 with probability $p$ and the value 0 with probability $(1 - p)$. We say $Y$ is a Bernoulli random variable and we have the following, where $E[\cdot]$ is used to denote expectation:

$$E[Y] = 1 * p + 0 * (1 - p) = p \tag{2.1}$$

In the continuous case, we instead weight by the probability density function. Notably, if $c$ is a constant, then $E[c] = c$ and the expectation of a mean 0 random variable is equal to 0. We will see the use of expectations in Section 2.2.3 and Chapter 5. Emphasis, however, is placed on the general concept of expectation as a weighted average, and technical details are not emphasized.

*Likelihood*

Maximum likelihood is probably the most widely used approach to finding point estimates of population-level parameters based on a sample of data. For the genetic association setting described herein, we may be interested, for example, in characterizing the effect of having a variant allele at a given SNP locus on a quantitative trait. Maximum likelihood is one approach to deriving an estimate of this effect based on a sample of observations that maintains several desirable properties. A complete introduction to estimation and associated concepts is provided in Casella and Berger (2002).

Briefly, suppose our parameter of interest is denoted $\theta$. The *likelihood function* is given by

$$L(\theta|\mathbf{y}) = L(\theta|y_1, \ldots, y_n) = f(y_1, \ldots, y_n|\theta) \tag{2.2}$$

where $n$ is the number of individuals in our sample and $f(y_1, \ldots, y_n|\theta)$ is the joint probability distribution of $\mathbf{y} = (y_1, \ldots, y_n)$. Under the assumption that our observations are independent and identically distributed, we have

$$L(\theta|\mathbf{y}) = \prod_{i=1}^{n} f(y_i|\theta) \qquad (2.3)$$

If we further assume our trait is normally distributed with mean $\mu$ and variance $\sigma^2$, then we have $f(y_i|\theta) = N(\mu, \sigma^2)$. Here $\theta$ is the vector $(\mu, \sigma^2)$. The *maximum likelihood estimate* of $\theta$, denoted $\widehat{\theta}$, is one that maximizes the likelihood function given in Equation (2.2) with respect to $\theta$. Notably, this is conditional on the observed data, $\mathbf{y}$. That is, the maximum likelihood estimate is the most probable value of the corresponding parameter, given the data at hand.

### 2.1.2 Important epidemiological concepts

In this section we review several notable epidemiological and biostatistical concepts, including confounding, effect mediation, interaction and conditional association, relevant to any population-based investigation. The importance of these concepts is highlighted throughout this text. For example, we see in Chapter 3 that race and ethnicity often play the role of a traditional confounder in the association between genotype and a trait. In Chapter 6, we discuss the distinction between interaction and conditional association when interpreting the results of a classification or regression tree.

*Confounding and effect mediation*

Consider the setting in which interest lies in characterizing the association between two variables, an exposure and an outcome. For example, we may want to determine whether heavy alcohol consumption (the exposure) is associated with total cholesterol level (the outcome). A *confounder* is defined as a variable that is: (1) associated with the exposure variable; (2) independently associated with the outcome variable; and (3) not in the causal pathway between exposure and disease. For example, smoking status is a potential confounder in the relationship above since smoking tends to be associated with heavy alcohol use and is also associated with cholesterol level among individuals who are not heavy alcohol users. In population-based genetic settings, we are generally interested in the association between genotype, as defined by one or more SNPs, and a trait. In this case, a confounder is defined as a clinical or demographic variable that is associated with both the genotype and trait under investigation, as illustrated in Figure 2.1.

In this figure, the solid, double-headed arrows are used to denote an association, while single-headed arrows are used to denote causality in the sense that one variable is believed to precede another. The dotted line between genotype and trait in this diagram is used to emphasize the fact that there may or may not be an association between these two variables. The existence of a true underlying association between the two variables of interest is irrelevant

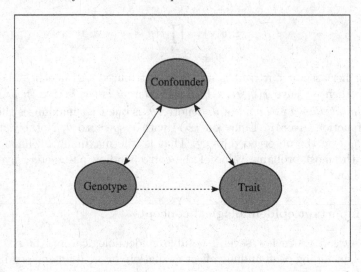

**Fig. 2.1.** Confounding
A confounder variable is associated with the genotype and independently associated
with the trait. We exclude the situation in which a third variable is in the causal
pathway between genotype and disease, as described by Figure 2.2.

to the definition a confounder. Importantly, however, ignoring the presence
of a confounder in the analysis can lead to erroneous conclusions about the
association of interest. Technically, since a confounder cannot be in the causal
pathway, the arrows cannot point both from the genotype to the confounder
and from the confounder to the trait.

In Section 3.1.3, we will see confounding in the context of assessing linkage disequilibrium (LD) in the presence of population stratification. LD is
defined as an association between the alleles on a single homologous chromosome between two genetic loci. If the allelic distribution at both of these
sites differs across subpopulations, such as racial or ethnic groups, then *population* serves as a confounder in the relationship between the two sites. In
the LD setting, we will see that combining data across subpopulations may
result in the appearance of LD when in fact there is no association between
the sites in either subpopulation considered alone. In population-based association studies, population admixture can similarly lead to the appearance
of an association between genotype and disease status when in fact such an
association does not exist. In this case, population serves to confound this
relationship if the likelihood of disease as well as the genotype frequencies
differ across subpopulations.

One common example of confounding is found in studies of cardiovascular disease that involve multiple racial or ethnic groups. As we describe in
Chapter 3, genotype frequencies often differ by race and ethnicity, and in
some cases the dominant allele at a given site even varies across these groups.

Several reports also suggest that lipid outcomes, such as total cholesterol and triglyceride levels, are associated with race and ethnicity, with Black non-Hispanics generally having a better lipid profile than Whites and Hispanics. This is a classic case of race and ethnicity confounding the relationship between genotype and lipid measures and may result in a spurious association between exposure and disease. Other clinical and demographic variables may also serve as confounders in genotype–trait association studies. For example, country of origin may be associated with both allele frequencies and a disease phenotype. Smoking and alcohol use are also thought to be associated with genetic polymorphisms and many disease traits. Careful consideration, however, is needed for variables such as these to distinguish between confounders and what are called effect mediators. This subtle yet very important distinction is discussed below. Further details on the challenges associated with different variable types in analysis are provided in Christenfeld *et al.* (2004). Identification of potential confounders will rely heavily on the input of the investigative team, inclusive of clinicians, epidemiologists, and geneticists.

A variable that lies in the causal pathway between the predictor and outcome of interest is called an *effect mediator* or causal pathway variable. A diagram illustrating the presence of an effect mediator in the association between genotype and a trait is provided in Figure 2.2. Here we see that the genotype affects the trait through alteration of the mediator variable. Consider for example a study that aims to determine whether there is an association between a single genetic polymorphism and lung cancer. If the polymorphism under study increases the likelihood that an individual smokes, and in turn it is the smoking that causes lung cancer, then we would say that smoking is an effect mediator in this case. By definition, a marker for disease is believed to be within the causal pathway to disease. Importantly, a thorough analysis plan must distinguish between patient-level covariates that serve as confounders and those that play the role of effect mediators. In general, while adjustment for confounding is essential in our analysis, inclusion of causal pathway variables is not recommended in the usual multivariable regression setting. Unfortunately, it is difficult to distinguish between confounding and effect mediation from an analytical standpoint. Scientific insight into the mechanisms of disease will elucidate these distinctions. Further discussion of the analytic challenges arising from confounding, effect mediation and the conflation of the two can be found in Robins and Greenland (1992), Rothman and Greenland (1998), Cole and Hernan (2002), Hernan *et al.* (2002) and Christenfeld *et al.* (2004).

*Interaction and conditional association*

*Effect modification* describes the setting in which the effect of a predictor variable on the outcome depends on the level of another variable, called a *modifier*. In this case, we say the predictor variable and the modifier *interact* in a statistical sense in their association with the outcome. Consider Figure 2.3, for example, where again we assume interest lies in characterizing the

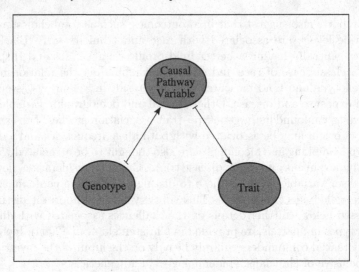

**Fig. 2.2.** Effect mediation
An effect mediator is in the causal pathway between the genotype and the trait. A confounder, on the other hand, is associated with the genotype and the trait, but is not in the causal pathway.

relationship between genotype and a trait. An effect modifier is defined as a variable that alters the relationship between the exposure and outcome. In this figure, the effect of genotype on the trait is given by $\beta_1$ when the effect modifier is equal to 0, while this effect is $\beta_2$ when the effect modifier is equal to 1. Formally, interaction is defined precisely as $\beta_1 \neq \beta_2$ and is assessed by testing the null hypothesis $H_0 : \beta_1 = \beta_2$. Rejection of this null implies statistical interaction. We return to a discussion of statistical interaction in the context of multivariable models in Section 2.2.3.

*Conditional association* is technically different from interaction, though the two terms are often used interchangeably. A conditional association is when the effect of $x$ on $y$ is statistically significant within either or both levels of a third variable, $z$. Returning to Figure 2.3, we say a conditional association exists if $\beta_1 \neq 0$ or $\beta_2 \neq 0$ (or both). That is, a test of conditional association is a test of the composite null, given by $H_0 : \beta_1 = 0$ and $\beta_2 = 0$. If either effect is different from 0, then we would reject this null in favor of a conditional association. Notably, a statistically significant conditional association between $x$ and $y$ (conditional on $z$) does not imply a significant statistical interaction between $x$ and $z$. Consider for example the extreme case in which $\beta_1 = \beta_2 = 4$. Assuming a reasonable spread and sample size, we expect to find that both $\beta_1 \neq 0$ and $\beta_2 \neq 0$ and reject the null; however, we do not expect to reject the interaction null ($H_0 : \beta_1 = \beta_2$). The difference between conditional association and interaction is an important one when considering methods that are designed specifically for detecting one or the

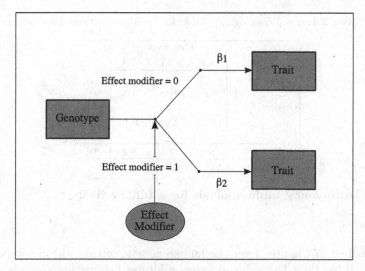

**Fig. 2.3.** Effect modification and conditional association

other. For example, we will see in Chapter 6 that classification and regression trees are formulated specifically for detecting conditional associations.

## 2.2 Measures and tests of association

This section provides a basic overview of a few simple statistical measures and tests of association. Several introductory biostatistics books cover these topics in greater detail, and the reader is referred to these for a comprehensive discussion; see for example Pagano and Gauvreau (2001) and Rosner (2006). As we will see below, the choice of statistical procedure used to assess association among variables depends on the hypothesis and structure of the data under investigation. Furthermore, estimation and testing can be either unadjusted or adjusted and univariate or multivariable. Adjusted analyses take into account potential confounders in the relationship between exposure (genotype in our setting) and the outcome. Multivariable methods involve multiple predictors and provide a venue for assessing interaction and conditional association. In the remainder of this chapter and Chapter 3, we use the term $p$-value without explicitly defining it, assuming the reader is broadly familiar with hypothesis testing. We define this term as well as other relevant testing concepts, such as type-1 and type-2 error rates, in Chapter 4.

**Table 2.1.** $2 \times 3$ contingency table for genotype–disease association

|  |  | Genotype | | |  |
| --- | --- | --- | --- | --- | --- |
|  |  | $aa$ | $Aa$ | $AA$ |  |
| Disease | $+$ | $n_{11}$ | $n_{12}$ | $n_{13}$ | $n_{1.}$ |
|  | $-$ | $n_{21}$ | $n_{22}$ | $n_{23}$ | $n_{2.}$ |
|  |  | $n_{.1}$ | $n_{.2}$ | $n_{.3}$ | $n$ |

## 2.2.1 Contingency table analysis for a binary trait

*Odds ratio*

As described in Chapter 1 for the human genetic setting, the genotype at a given SNP has three levels: homozygous wildtype, heterozygous, and homozygous rare. If the outcome is binary, then the data can be represented by the $2 \times 3$ contingency table given in Table 2.1. Here $n_{ij}$ is the number of individuals in the corresponding cell for $i = 1, 2$ and $j = 1, 2, 3$. For example, $n_{11}$ is the number of individuals in our sample who have disease and express the $aa$ genotype. In this setting, a commonly used measure of association is the odds ratio (OR), defined as the ratio of the odds of disease among the exposed to the odds of disease among the unexposed. In our setting, genotype can be thought of as the *exposure*, so that the OR is the ratio of the odds of disease given a specific genotype to the odds of disease among individuals without the specified genotype.

Formally, if $D^+$ and $D^-$ represent diseased and not diseased and $E^+$ and $E^-$ represent exposed and not exposed, respectively, then the OR is written

$$OR = \frac{Pr(D^+|E^+)/[1 - Pr(D^+|E^+)]}{Pr(D^+|E^-)/[1 - Pr(D^+|E^-)]} \qquad (2.4)$$

where | indicates conditional, as described in Section 2.1.1. In the case of a three-level exposure variable, it is common to calculate the OR for each level of exposure in relation to the referent group. In the genetics setting, we often calculate the OR for each genotype in relation to the homozygous wildtype genotype, $AA$. That is, we define $OR_{aa,AA}$ as the odds of disease for individuals with the homozygous rare genotype, $aa$, compared with those with the homozygous wildtype genotype $AA$. Likewise $OR_{Aa,AA}$ is the odds of disease for heterozygous individuals compared with those individuals with the $AA$ genotype. Algebraically, based on Equation (2.4) and the notation in Table 2.1, we have

$$OR_{aa,AA} = \frac{(n_{11}/n_{.1})/(n_{21}/n_{.1})}{(n_{13}/n_{.3})/(n_{23}/n_{.3})} = \frac{n_{11}n_{23}}{n_{21}n_{13}} \qquad (2.5)$$

**Table 2.2.** $2 \times 2$ contingency table for genotype–disease association

|  |  | Genotype | |  |
|---|---|---|---|---|
|  |  | $(Aa$ or $aa)$ | $AA$ |  |
| Disease (D) | $+$ | $n_{11}$ | $n_{12}$ | $n_{1.}$ |
|  | $-$ | $n_{21}$ | $n_{22}$ | $n_{2.}$ |
|  |  | $n_{.1}$ | $n_{.2}$ | $n$ |

since $Pr(D^+|E^+) = n_{11}/n_{.1}$ and $Pr(D^+|E^-) = n_{13}/n_{.3}$. Similarly, it is straightforward to show that $OR_{Aa,AA} = (n_{12}n_{23})/(n_{22}n_{13})$.

One alternative approach is to dichotomize genotype prior to calculating the OR. For example, we can define genotype as an indicator for the presence of any variant allele. Suppose the three possible genotypes are again given by $AA$, $Aa$, and $aa$. Then a dichotomized genotype exposure could be defined as $E^+ = (Aa$ or $aa)$ and $E^- = (AA)$. The corresponding count data are now given by Table 2.2. In this case, we have a $2 \times 2$ contingency table, and the single corresponding OR is equal to

$$\widehat{OR} = \frac{(n_{11}/n_{.1})/(n_{21}/n_{.1})}{(n_{12}/n_{.2})/(n_{22}/n_{.2})} = \frac{n_{11}n_{22}}{n_{21}n_{12}} \tag{2.6}$$

An estimate of the OR can be calculated in R using the `oddsratio()` function in the `epitools` package.

### Pearson's $\chi^2$-test and Fisher's exact test

A formal test of association between a categorical exposure (genotype) and categorical disease variable (trait) is conducted using Pearson's $\chi^2$-test or Fisher's exact test. In the context of a $2 \times 2$ table, a test of no association between the rows and columns is equivalent to a test of the single null hypothesis, $H_0 : OR = 1$. Pearson's $\chi^2$-test involves first determining the expected cell counts of a corresponding contingency table under the assumption of independence between the genotype and trait. These expected counts are calculated using the concept of independence described in Section 2.1.1. Recall that the probability of two independent events, in this case exposure and disease, is simply the product of the probabilities of each event. Thus, for example, the probability that an individual contributes to the $(1,1)$-cell, under independence, is given by $Pr(D^+)Pr(E^+) = (n_{1.}/n)(n_{.1}/n) = n_{1.}n_{.1}/n^2$. We multiply this by $n$, the total number of individuals in our sample, to get a corresponding expected count, $E_{11} = n_{1.}n_{.1}/n$.

More generally, the expected count for the $(i,j)$-cell is given by $E_{ij} = n_{i.}n_{.j}/n$ for $i = 1, 2$ and $j = 1, 2, 3$. Letting the corresponding observed cell counts be denoted $O_{ij}$, Pearson's $\chi^2$-statistic is given by

$$\chi^2 = \sum_{i,j} \frac{(O_{ij} - E_{ij})^2}{E_{ij}} \sim \chi^2_{(r-1)(c-1)} \qquad (2.7)$$

This statistic has a $\chi^2$-distribution with $(r-1)(c-1)$ degrees of freedom, where $r = 2$ and $c = 3$ are the number of rows and columns, respectively. Application of this testing framework is straightforward in R using the `chisq.test()` and is described in the example below. Further application of this approach is presented in the context of assessing association between alleles for two homologous chromosomes (that is, for testing Hardy-Weinberg equilibrium (HWE)) in Chapter 3.

*Example 2.1 (Chi-squared test for association).* Suppose we are interested in determining whether there is an association between any of the SNPs within the `esr1` gene and an indicator of body mass index (BMI) $> 25$ at baseline of the FAMuSS study. We use the following code first to identify the names of all of the `esr1` SNPs:

```
> attach(fms)
> NamesEsr1Snps <- names(fms)[substr(names(fms),1,4)=="esr1"]
> NamesEsr1Snps
```

```
[1] "esr1_rs1801132" "esr1_rs1042717" "esr1_rs2228480" "esr1_rs2077647"
[5] "esr1_rs9340799" "esr1_rs2234693"
```

The genotype matrix can now be defined by selecting the columns of `fms` that correspond to the `esr1` SNP names:

```
> fmsEsr1 <- fms[,is.element(names(fms),NamesEsr1Snps)]
```

We define our trait to be an indicator for whether BMI is $> 25$ at baseline:

```
> Trait <- as.numeric(pre.BMI>25)
```

Next we write a function to record the $p$-values from applying the $\chi^2$-test to the $2 \times 3$ contingency tables corresponding to each SNP and this trait:

```
> newFunction <- function(Geno){
>         ObsTab <- table(Trait,Geno)
>         return(chisq.test(ObsTab)$p.value)
>         }
```

Finally, we apply this function to the columns of `fmsEsr1`:

```
> apply(fmsEsr1,2,newFunction)
```

```
[1] 0.4440720 0.0264659 0.1849870 0.1802880 0.1606800 0.1675418
```

Based on this output, we see the suggestion of an association between the second SNP, `esr1_rs1042717`, and BMI. A closer look at these data yields

```
> Geno <- fmsEsr1[,2]
> ObsTab <- table(Trait,Geno)
> ObsTab

    Geno
y    AA  GA  GG
  0  30 246 380
  1  30 130 184
```

This reveals that individuals with the $AA$ genotype are equally likely to have a BMI greater than or less than 25, while the proportion of individuals with the $GG$ genotype who have a BMI greater than 25 is only $184/(380 + 184) = 32.6\%$. Importantly, this analysis does not correct for multiple testing, which, as discussed in Section 2.3.1, leads to an inflation of the type-1 error rate; that is, the probability of rejecting the null hypothesis when it is in fact true. We return to this example and consider appropriate multiple comparison adjustments in Chapter 4. □

Fisher's exact test is preferable when at least 20% of the expected cell counts are small ($E_{ij} < 5$). The exact $p$-value is given by the probability of seeing something as extreme or more extreme in the direction of the alternative hypothesis than what is observed. Fisher derived this probability for the $2 \times 2$ table of Table 2.2, and it is defined explicitly in Section 3.2.1 for testing HWE. Implementation of this test of association is straightforward using the fisher.test() function in R, as demonstrated in the following example.

*Example 2.2 (Fisher's exact test for association).* Returning to Example 2.1, suppose we are again interested in determining whether there is an association between any of the SNPs within the esr1 gene and an indicator of body mass index (BMI) > 25 at baseline of the FAMuSS study. In this case, we create a function that returns Fisher's exact $p$-value for a test of association between each of the SNPs and the trait using the Trait variable created in Example 2.1:

```
> newFunction <- function(Geno){
>        ObsTab <- table(Trait,Geno)
>        return(fisher.test(ObsTab)$p.value)
>        }
```

Application of this function to the fmsEsr1 matrix of genotypes, created in Example 2.1, yields

```
> apply(fmsEsr1,2,newFunction)

[1] 0.46053113 0.02940733 0.18684765 0.17622428 0.15896064 0.16945636
```

In this case, the $p$-values of the two approaches to analysis (Fisher's exact test and Pearson's $\chi^2$-test) are comparable since the asymptotic assumptions of the $\chi^2$-test are not violated. In both cases, there is the suggestion of an association between the second SNP, esr1_rs1042717, and BMI. □

*Correlation*

The term *correlation* is often used in a general sense to refer to an association between two variables. For example, alcohol use and smoking status are often said to be "correlated" since smokers tend to drink more alcohol than non-smokers. The *correlation coefficient* between two random variables is defined mathematically as the ratio of the covariance between these two variables and the product of their respective standard deviations. So defined, the correlation coefficient is a measure of linear association between two variables and takes on values between $-1$ and $1$. The two most commonly used sample-based estimates of the correlation coefficient are Pearson's product-moment correlation coefficient and Spearman's rank correlation coefficient. These are often referred to as the Pearson and Spearman coefficients, respectively, and are appropriate measures for continuous variables.

A primary limitation of Pearson's coefficient is that it is highly sensitive to outlying values. In these settings, the non-parametric alternative, Spearman's rank correlation coefficient, is generally preferred. In the genetic association setting, we are often interested in relating a categorical genotype variable to a quantitative or binary trait. The correlation between two dichotomous variables for which the underlying distribution is not assumed to have continuity is estimated by the *phi-coefficient*, denoted $\phi$. This is calculated using the same formula as Pearson's correlation coefficient and is closely related to the $\chi^2$-statistic for a test of association between two binary variables. In fact, it can be shown that $\phi = \sqrt{r^2}$, where $r^2$ is defined in Section 3.1.1 as a measure of association, also referred to as linkage disequilibrium (LD), between two genetic loci.

*Cochran-Armitage (C-A) trend test*

The Cochran-Armitage (C-A) trend test is commonly applied in the analysis of association between a biallelic locus, with three corresponding genotype levels, and a binary trait. The C-A trend test is a test developed specifically for detecting a *linear trend* in proportions over levels of the exposure variable. In our setting, it is a test of whether the probability of disease increases linearly with genotype. We begin by taking into account the ordering of the columns of the contingency table. For example, we may choose to code the genotype data as 0, 1, or 2 for the number of $A$ alleles. If indeed the relationship between genotype and disease is linear, the C-A test will be more powerful than Pearson's $\chi^2$-test. Formally, returning to Table 2.1, let $p_j$ be the probability of disease for the $j$th genotype column. A trend test can be defined as a test of the null hypothesis that there is no linear relationship among the $p_j$'s. A linear relationship is expressed algebraically as $p_j = \alpha + \beta S_j$, where $S_j$ is a score for the $j$th column. For example, we can define $S_1 = 1$, $S_2 = 2$ and $S_3 = 3$. Thus, the null hypothesis of no linear association is given by $H_0 : \beta = 0$. Testing of this hypothesis is achieved using a $\chi^2$-test statistic that is a function of the observed and expected scores and is illustrated in the following example.

*Example 2.3 (Cochran-Armitage (C-A) trend test for association).* In this example, we demonstrate application of the C-A trend test for association between the esr1_rs1042717 SNP and baseline BMI for the FAMuSS data, where again BMI is defined as a binary indicator for BMI $> 25$. This test statistic can be calculated using the independence_test() function in the coin package in R. We begin by installing several packages required by the coin package, including splines, survival, stats4, mvtnorm and modeltools, and then call the coin package:

```
> install.packages("coin")
> library(coin)
```

We then define our genotype and trait variables without needing to exclude observations with missing genotypes:

```
> attach(fms)
> Geno <- esr1_rs1042717
> Trait <- as.numeric(pre.BMI>25)
```

The independence_test() function, with the option testat="quad" specified, can then be applied to conduct the C-A trend test. Here we use the ordered() function to specify that the genotypes are ordered. The scores option allows us to specify the relationship between different levels of our genotype:

```
> GenoOrd <- ordered(Geno)
> independence_test(Trait~GenoOrd,teststat="quad",
+         scores=list(GenoOrd=c(0,1,2)))

        Asymptotic General Independence Test

data:  Trait by Geno.or (AA < GA < GG)
chi-squared = 4.4921, df = 1, p-value = 0.03405
```

In this case, the $p$-value is greater than we saw for the corresponding SNP in Example 2.1. Returning to that example, we saw that the proportion of individuals with BMI greater than 25 does not decrease linearly with genotype at this SNP. In fact, these proportions are 0.50, 0.35 and 0.33 for 2, 1 and 0 copies of the $A$ allele, respectively. □

Methods for adjusting for a single categorical confounder or effect modifier are well described in the context of contingency table analysis, see for example Rosner (2006) and Agresti (2002). The more general linear and logistic regression frameworks, described in Section 2.2.3, offer an alternative setting for analysis. The primary advantage of the regression settings is that they provide for simultaneous consideration of multiple potential confounders and effect modifiers. Furthermore, the regression framework is a natural setting for consideration of both categorical and continuous covariates. Before presenting this setting, we offer a brief description of simple tests of association for quantitative traits.

### 2.2.2 M-sample tests for a quantitative trait

In the previous section, we focused on a binary trait such as disease status and described methods for assessing genotype–trait associations in that setting. Now we turn to consideration of quantitative traits and discuss analytic methods for the setting in which we aim to characterize the association between genotype and this quantitative trait. Genotype can be defined as an M-level factor and in the simplest case reduces to a binary indicator, for example, for the presence of at least one variant allele at a given SNP locus. We begin by describing the *two-sample t-test* and *Wilcoxon rank-sum test*, which are typically applied to test for association in this setting. We then present the analysis of variance (ANOVA) and the non-parametric analog, the Kruskal-Wallis (K-W) test, which are applicable in the more general context of multiple genotype levels.

*Two-sample t-test and Wilcoxon rank-sum test*

Formally, the *t*-test is a test of the null hypothesis that the mean is the same in two populations, written $H_0 : \mu_1 = \mu_2$, where in our setting the populations are defined based on genotype. For example, we might define $\mu_1$ as the population mean for individuals with the $AA$ genotype and $\mu_2$ as the population mean for individuals with the $Aa$ or $aa$ genotypes. The two-sample *t*-test statistic, assuming equal variances, is given by

$$t = \frac{\bar{y}_1 - \bar{y}_2}{\sqrt{s_p^2 \left[1/n_1 + 1/n_2\right]}} \sim T_{n_1+n_2-2} \tag{2.8}$$

where $\bar{y}_1$ and $\bar{y}_2$ are the sample means of the quantitative trait for genotype groups 1 and 2, $s_p^2$ is the pooled estimate of the variance, and $n_1$ and $n_2$ are the respective sample sizes. Under the null, this statistic has a $T$-distribution with degrees of freedom equal to $n_1 + n_2 - 2$.

The Wilcoxon rank-sum test (also called the Mann-Whitney $U$-test) is a non-parametric analog to the two-sample *t*-test and is more appropriate than the *t*-test if the trait is not normally distributed and the sample size is small. The Wilcoxon rank-sum test is a rank-based test and is used to test the null hypothesis that the medians of a quantitative trait in each of two populations are equal. Both the *t*-test and Wilcoxon rank-sum test are easily implemented in R using the `t.test()` and `wilcox.test()` functions, respectively, as illustrated in the following example.

*Example 2.4 (Two-sample tests of association for a quantitative trait).* Returning to the FAMuSS data, suppose we are interested in determining whether having at least one copy of the variant allele for any of the SNPs within the `resistin` gene is associated with a change in the non-dominant muscle strength before and after exercise training, given by the variable `NDRM.CH`. In this case, we begin by creating a vector of names for the SNPs within the `resistin` gene and a corresponding matrix of genotypes.

```
> attach(fms)
> NamesResistinSnps <- names(fms)[substr(names(fms),1,8)=="resistin"]
> fmsResistin <- fms[,is.element(names(fms),NamesResistinSnps)]
```

We then create a new function that takes a single genotype vector, converts it to binary elements, and generates $p$-values based on the $t$-test of equality of the mean trait across the resulting two levels. We first call the `genetics` package since we are using the `allele.names()` function within this package:

```
> library(genetics)
> TtestPval <- function(Geno){
+        alleleMajor <- allele.names(genotype(Geno, sep="",
+            reorder="freq"))[1]
+        GenoWt <- paste(alleleMajor, alleleMajor, sep="")
+        GenoBin <- as.numeric(Geno!=GenoWt)[!is.na(Geno)]
+        Trait <- NDRM.CH[!is.na(Geno)]
+        return(t.test(Trait[GenoBin==1], Trait[GenoBin==0])$p.value)
+    }
```

Here we define the binary genotype variable (`GenoBin`) as an indicator for at least one variant allele at the corresponding site. Notably, this amounts specifically to testing a dominant genetic model in which one or two copies of the variant allele result in a shift in the mean of `NDRM.CH` compared with the homozygous wildtype genotype, as we will see in Section 2.3.4.

Next we apply this function, `TtestPval`, to each of the columns of the genotype matrix, `fmsResistin`:

```
> apply(fmsResistin,2,TtestPval)

 resistin_c30t resistin_c398t resistin_g540a resistin_c980g
    0.04401614     0.08098567     0.11578470     0.27828906
resistin_c180g resistin_a537c
    0.03969448     0.06573061
```

This analysis suggests that the first and fifth SNPs within `resistin` may be associated with a change in NDRM before and after exercise training; however, we have not adjusted for multiple testing, and our conclusions are not yet decisive. Closer examination of the output from application of the `t.test()` function provides additional information on the nature of the underlying association:

```
> Geno <- fms$"resistin_c180g"
> table(Geno)

Geno
 CC  CG  GG
330 320  89

> GenoWt <- names(table(Geno))[table(Geno)==max(table(Geno))]
> GenoWt
```

```
[1] "CC"

> GenoBin <- as.numeric(Geno!=GenoWt)[!is.na(Geno)]
> Trait <- NDRM.CH[!is.na(Geno)]
> t.test(Trait[GenoBin==1],Trait[GenoBin==0])

        Welch Two Sample t-test

data:  Trait[GenoBin == 1] and Trait[GenoBin == 0]
t = -2.0618, df = 552.158, p-value = 0.03969
alternative hypothesis: true difference in means is not equal to 0
95 percent confidence interval:
 -10.9729548  -0.2658088
sample estimates:
mean of x mean of y
 50.43503  56.05441
```

The output reveals that the sample mean change in NDRM is 56.05 among individuals with the CC genotype while the mean change is 50.44 among individuals with at least one variant allele at this site (CG or GG). Thus, the presence of at least one variant allele appears to lead to a decrease in the change in NDRM before and after exercise training. Again, an adjustment for multiple comparisons is needed to make firm conclusions, as described in Chapter 4. A similar approach can be taken to apply the Wilcoxon rank-sum test, where the t.test() function is replaced with the wilcox.test() function.                                                                    □

*Analysis of variance (ANOVA) and Kruskal-Wallis (K-W) test*

If *a priori* dichotomization of the genotype variables is not desirable, we can perform an analysis of variance (ANOVA) or the non-parametric analog, the Kruskal-Wallis test, to characterize association with a quantitative trait. ANOVA is an extension of the two-sample $t$-test to the $M$-sample setting and is based on an $F$-test for a full model with $M-1$ genotype indicators (dummy variables) compared with the reduced model with an overall mean. The Kruskal-Wallis test similarly extends the Wilcoxon rank-sum test. Applications of these tests are illustrated in the following example. Further details on ANOVA and its relevance to the analysis of data arising from genetic association studies are provided in Section 4.4.2.

*Example 2.5 (M-sample tests of association for a quantitative trait).* Suppose we are again interested in determining whether there is an association between the resistin_c180g SNP and the percentage change in the non-dominant arm muscle strength before and after exercise training, as measured by NDRM.CH, based on the FAMuSS data. Further, suppose we do not want to impose any prior assumptions about the genetic model and so decide to treat genotype as a three-level factor variable. We begin by reading in the relevant genotype data as a factor variable and defining the trait:

```
> attach(fms)
> Geno <- as.factor(resistin_c180g)
> Trait <- NDRM.CH
```

We use the lm() function in R to perform ANOVA. Alternatively, the aov() function can be applied, though the corresponding output of the associated print() and summary() functions will differ. Here we indicate that we want to exclude individuals with missing values for the trait, coded as NA, using the option na.action==na.exclude. The summary() function takes as input an object of class "lm" and provides us with the details of the model-fitting results, including the overall $F$-test for association.

```
> AnovaMod <- lm(Trait~Geno, na.action=na.exclude)
> summary(AnovaMod)

Call:
lm(formula = Trait ~ Geno, na.action = na.exclude)

Residuals:
    Min      1Q  Median      3Q     Max
-56.054 -22.754  -6.054  15.346 193.946

Coefficients:
            Estimate Std. Error t value Pr(>|t|)
(Intercept)   56.054      2.004  27.973   <2e-16 ***
GenoCG        -5.918      2.864  -2.067   0.0392 *
GenoGG        -4.553      4.356  -1.045   0.2964
---
Signif. codes:  0 *** 0.001 ** 0.01 * 0.05 . 0.1   1

Residual standard error: 33.05 on 603 degrees of freedom
  (791 observations deleted due to missingness)
Multiple R-squared: 0.007296,   Adjusted R-squared: 0.004003
F-statistic: 2.216 on 2 and 603 DF,  p-value: 0.1100
```

This analysis yields an overall $F$-test statistic of $F_{2,603} = 2.216$ ($p = 0.11$). Based on this model and the data at hand, we therefore conclude that there is not enough evidence to suggest an association between resistin_c180g and NDRM.CH. Interestingly, based on a Wald test, there appears to be an effect of the CG genotype compared with the referent CC genotype ($t = -2.067$, $p = 0.0392$); however, interpretation of this $p$-value needs to be in light of the two $t$-statistics generated unless there was an *a priori* hypothesis about the heterozygous genotype in this setting. Further discussion of the appropriate adjustment is given in Chapter 4.

A Kruskal-Wallis (K-W) test can also be applied and is more appropriate in small-sample settings in which the assumption of normality may not be reasonable. This is straightforward using the following code. Again we specify na.action==na.exclude to exclude individuals with missing values for the trait, which are coded as NA:

```
> kruskal.test(Trait, Geno, na.action=na.exclude)

        Kruskal-Wallis rank sum test

data:  Trait and Geno
Kruskal-Wallis chi-squared = 4.9268, df = 2, p-value = 0.08515
```

In this example, the K-W test yields a $p$-value of 0.085, and a similar conclusion is reached that there is not sufficient evidence of an association between resistin_c180g and percentage change in muscle strength of the non-dominant arm as measured by NDRM.CH. □

## 2.2.3 Generalized linear model

In the previous sections, we focused on tests of association between genotype and a trait without specific consideration of additional covariates. These univariate methods can easily be applied within strata (levels) of a third variable. Such a stratified analysis is common if there is reason to believe that the effect of genotype on the trait varies depending on the value of another variable. For example, we may stratify by smoking status if we think the effect of a variant allele on lipid abnormalities is potentially different for smokers and non-smokers. Alternatively, we can fit a multivariable model to either quantitative or binary traits. Multivariable models have the primary advantage of allowing us to account appropriately for multiple potential confounders and effect modifiers. Several excellent texts provide an introduction to linear modeling; see for example Faraway (2005) and Neter *et al.* (1996). Here we offer a brief summary of important concepts used throughout this text.

The generalized linear model (GLM) is given in matrix notation by the equation

$$g(E[\mathbf{y}]) = \mathbf{X}\beta \tag{2.9}$$

where $E[\mathbf{Y}] = \mu$ denotes the expectation of $\mathbf{Y}$, as described in Section 2.1.1, $g()$ is a link function and $\mathbf{X}$ is the design matrix. The generalized linear model should not be confused with the general linear model for multivariate data. The generalized linear model, the topic of this section, is a modeling framework that is applicable to a variety of dependent variables, including both quantitative and binary traits, as well as count data. Here we begin by describing the linear regression model for quantitative traits and then introduce the logistic model for binary outcomes. Both represent special cases of the generalized linear model.

*Simple and multivariable linear regression*

In the case of a quantitative trait, we let $g()$ be the identity link, and Equation (2.9) reduces to the ordinary linear regression model. That is, we have

$$g(E[\mathbf{y}]) = E[\mathbf{y}] = \mathbf{X}\boldsymbol{\beta} \tag{2.10}$$

or equivalently

$$\mathbf{y} = \mathbf{X}\boldsymbol{\beta} + \boldsymbol{\epsilon} \tag{2.11}$$

For example, a simple linear regression model for a quantitative trait is given by Equation (2.11) where

$$\mathbf{y} = \begin{bmatrix} y_1 \\ y_2 \\ \vdots \\ y_n \end{bmatrix} ; \quad \mathbf{X} = \begin{bmatrix} 1 & x_1 \\ 1 & x_2 \\ \vdots \\ 1 & x_n \end{bmatrix} ; \quad \boldsymbol{\epsilon} = \begin{bmatrix} \epsilon_1 \\ \epsilon_2 \\ \vdots \\ \epsilon_n \end{bmatrix} \tag{2.12}$$

and $\boldsymbol{\beta} = (\beta_0, \beta_1)^T$. This model includes an intercept, represented by the first column of $\mathbf{X}$, and a single predictor variable, given by the second column of $\mathbf{X}$. In general, we assume the error terms, $\epsilon_i$ for $i = 1, \ldots, n$, are independent and identically distributed with mean 0.

The scalar formulation of this simple linear regression model is given by

$$y_i = \beta_0 + \beta_1 x_i + \epsilon_i \tag{2.13}$$

where $i = 1, \ldots, n$ indicates individual. In this model, the measure of association between $x$ and $y$ is given by the parameter $\beta_1$, defined as the amount of change in $y$ that occurs with one unit change in $x$. For example, if $x$ is an indicator for the presence of a variant allele at a given SNP locus and $y$ is cholesterol level, then $\beta_1$ is the difference in mean cholesterol level between individuals with and without this variant allele. The least squares estimates of $\beta_0$, the overall mean, and $\beta_1$, the association parameter, are given respectively by

$$\widehat{\beta}_0 = \left( \sum_i y_i - \widehat{\beta}_1 \sum_i x_i \right) / n \tag{2.14}$$

and

$$\widehat{\beta}_1 = \frac{n \sum_i x_i y_i - \sum_i x_i \sum_i y_i}{n \sum_i x_i^2 - \left( \sum_i x_i \right)^2} \tag{2.15}$$

Notably, the estimates above do not require an assumption of normality of the error terms. Importantly, $\beta$ is a measure of *linear* association, capturing information on the extent to which the relationship between $x$ and $y$ is a straight line.

The multivariable linear model is a generalization of the model given in Equation (2.13) in which additional variables are included on the right-hand side of the equation. For example, suppose we have $m$ covariates, given by $z_{i1}, \ldots, z_{im}$ for the $i$th individual. For example, $z_{i1}$ may be gender and $z_{i2}$ may represent smoking status for individual $i$. We are often interested in fitting a model of the form

$$y_i = \beta_0 + \beta_1 x_i + \sum_{j=1}^{m} \alpha_j z_{ij} + \epsilon_i \tag{2.16}$$

We call this a multivariable linear regression model since multiple independent variables are included. Again the measure of association between the genotype and trait is given by $\beta_1$. Now, however, estimation and testing of this parameter takes into account the additional variables in the model. These additional variables may be confounders or may help to explain the variability in our trait. As described in Section 2.1.2, the inclusion of confounding variables is important for drawing valid conclusions about the effect of genotype on the trait. Adding non-confounding variables to the model will not change our genotype effect estimate substantially. However, by reducing the unexplained variability in our model, including these variables may increase our power to detect the association of our primary independent variable.

A multivariable regression model is also a natural framework for consideration of effect modification by continuous or categorical variables. Suppose for example we are engaged in a pharmacogenomic investigation where interest lies in characterizing the extent to which the effect of a drug exposure on our trait varies by genotype. Statistically, we aim to estimate the interaction between genotype and drug exposure on our trait. Let genotype again be represented by $x$, the drug exposure given by $z$ and a quantitative trait measured by $y$. In this case, we construct an interaction model given by

$$y_i = \beta_0 + \beta_1 x_i + \beta_2 z_i + \gamma x_i z_i + \epsilon_i \tag{2.17}$$

Here $\gamma$ is the interaction effect and represents the additional effect of $z$ when the genotype is present.

For example, suppose $x$ is an indicator for at least one polymorphism in ApoCIII, a gene that has previously been associated with triglyceride levels. Further, suppose $z$ is an indicator for current exposure to lipid-lowering therapy (LLT) and $y$ is fasting triglyceride level. According to the model given in Equation (2.17), the effect of LLT on triglycerides is $\beta_2$ among individuals with no polymorphisms in ApoCIII ($x_i = 0$) and $\beta_2 + \gamma$ among individuals with at least one polymorphism in ApoCIII ($x_i = 1$). In other words, the interaction effect, $\gamma$, is the difference between the effect of $z$ on $y$ among those individuals with a polymorphism and the effect of $z$ on $y$ among those without a polymorphism. This model can also be represented in matrix notation by Equation (2.11), where now the design matrix, $\mathbf{X}$, is defined as

$$\mathbf{X} = \begin{bmatrix} 1 & x_1 & z_1 & (x_1 \times z_1) \\ 1 & x_2 & z_2 & (x_2 \times z_2) \\ \vdots & & & \\ 1 & x_n & z_n & (x_n \times z_n) \end{bmatrix} \qquad (2.18)$$

and $\boldsymbol{\beta} = (\beta_0, \beta_1, \beta_2, \gamma)^T$.

It is important to distinguish between interaction and a multiplicative effect. Consider for example the setting in which we take a natural log transformation of our trait, $y$, in order to normalize the data. In this case, an additive linear regression model is written

$$\ln(y_i) = \beta_0 + \beta_1 x_i + \beta_2 z_i + \epsilon_i \qquad (2.19)$$

or equivalently

$$y_i = \exp[\beta_0] \exp[\beta_1 x_i] \exp[\beta_2 z_i] \exp[\epsilon_i] \qquad (2.20)$$

Here we see that the effects of $x$ and $z$ on $y$ are multiplicative since the effect of a unit change in $x$ is an $\exp[\beta_1]$-fold increase in $y$ and likewise the effect of a unit change in $z$ is an $\exp[\beta_2]$-fold increase in $y$. The effect of $x$ on $y$, however, does not depend on the level of $z$. That is, whether $z_i = 0$ or $z_i = 1$, a unit change in $x$ results in the same increase in $y$. Similarly, the effect of $z$ on $y$ does not depend on the level of $x$. Therefore, this is a multiplicative model but not an interaction model.

In all of the modeling schemes above, interest lies both in estimating the model parameters and in testing hypotheses relating to their true population values. For example, for the additive model given in Equation (2.16), we may be interested in testing the null hypothesis of no association between the genotype and trait, given by $H_0 : \beta_1 = 0$. Testing the null hypothesis that there is no statistical interaction, on the other hand, will be based on the model in Equation (2.17) and given by $H_0 : \gamma = 0$. Notably, in traditional statistical applications, we do not test hypotheses that involve only main effects of a variable if indeed an interaction exists between that variable and another variable. That is, for the model in Equation (2.17), we would not test the hypothesis that $\beta_1 = 0$ unless we first conclude that there is no evidence for interaction, such that $\gamma = 0$.

Application of the Wald test or likelihood ratio test (LRT), which is equivalent to the $F$-test, is reasonable for testing hypotheses about parameters in the multivariable regression setting. The LRT makes few asymptotic assumptions and may be preferable if the results differ between methods. Further details on these testing procedures can be found in a number of intermediate biostatistics texts, including McCulloch and Searle (2001), Faraway (2005) and Neter *et al.* (1996). Importantly, while least squares estimation of the model parameters does not require a normality assumption on the errors, inference procedures for testing hypotheses based on the linear regression model generally assume the trait is normally distributed. Fitting a multivariable linear model is straightforward using the lm() function in R, as demonstrated in the following example.

*Example 2.6 (Linear regression).* In this example, we consider the association of the actn3_r577x SNP with the percentage change in strength of the non-dominant arm before and after exercise training based on the FAMuSS data. Our investigative team believes the effect of genotype on our trait may be modified by gender, and so we decide to fit a multivariable model with an interaction term. First we define our genotype and trait variables:

```
> attach(fms)
> Geno <- actn3_r577x
> Trait <- NDRM.CH
```

The full model includes terms for genotype, gender and a genotype by gender interaction:

```
> ModFull <- lm(Trait~Geno+Gender+Geno*Gender, na.action=na.exclude)
> summary(ModFull)

Call:
lm(formula=Trait ~ Geno + Gender + Geno * Gender, na.action=na.exclude)

Residuals:
    Min      1Q  Median      3Q     Max
-68.778 -20.258  -4.351  14.970 184.970

Coefficients:
                  Estimate Std. Error t value Pr(>|t|)
(Intercept)         53.558      2.964  18.068  < 2e-16 ***
GenoCT              11.471      3.882   2.955 0.003251 **
GenoTT              15.219      4.286   3.551 0.000414 ***
GenderMale         -13.507      4.769  -2.832 0.004777 **
GenoCT:GenderMale  -14.251      6.117  -2.330 0.020160 *
GenoTT:GenderMale  -13.537      6.867  -1.971 0.049153 *
---
Signif. codes:  0 *** 0.001 ** 0.01 * 0.05 . 0.1   1

Residual standard error: 30.8 on 597 degrees of freedom
  (794 observations deleted due to missingness)
Multiple R-squared: 0.1423, Adjusted R-squared: 0.1351
F-statistic: 19.81 on 5 and 597 DF,  p-value: < 2.2e-16
```

We compare this model to the reduced model without an interaction term by first fitting the reduced model:

```
> ModReduced <- lm(Trait~Geno+Gender, na.action=na.exclude)
```

The `anova()` function in R provides us with an $F$-test comparing the full and reduced models. Formally, this is a test of the null hypothesis that there is no interaction between genotype and gender:

```
> anova(ModReduced, ModFull)

Analysis of Variance Table

Model 1: Trait ~ Geno + Gender
Model 2: Trait ~ Geno + Gender + Geno * Gender
  Res.Df    RSS Df Sum of Sq      F  Pr(>F)
1    599 572345
2    597 566518  2      5827 3.0702 0.04715 *
---
Signif. codes:  0 *** 0.001 ** 0.01 * 0.05 . 0.1  1
```

The $F$-test statistic with 2 and 597 degrees of freedom is 3.07, with a corresponding $p$-value of 0.047. We therefore conclude that there does appear to be an interaction between gender and genotype based on these data and under the modeling assumptions. Based on the full model results above, we conclude that the overall percentage change in the non-dominant arm muscle strength is 53.56 for females with the homozygous wildtype CC genotype. In females, the effect of having the heterozygous CT genotype compared with CC is 11.47 and the effect of having the homozygous variant TT genotype in females is 15.22. In males, on the other hand, the effect on NDRM.CH of the CT genotype is $11.47 - 14.25 = -2.78$ and the effect of carrying the TT genotype is $15.22 - 13.54 = 1.68$.

Finally, the predicted change in muscle strength for the non-dominant arm and corresponding prediction intervals based on the full model and for each level of genotype and gender can be generated as follows:

```
> NewDat <- data.frame(Geno=rep(c("CC","CT","TT"),2),
+       Gender=c(rep("Female",3), rep("Male",3)))
> predict.lm(ModFull, NewDat, interval="prediction", level=0.95)

       fit        lwr       upr
1 53.55833  -7.220257 114.33692
2 65.02980   4.330657 125.72895
3 68.77778   7.973847 129.58171
4 40.05147 -20.890899 100.99384
5 37.27168 -23.494571  98.03793
6 41.73437 -19.235589 102.70434
```

From this output, we see that the predicted change in females increases steadily from 53.56 to 65.03 to 68.78 for the CC, CT and TT genotypes. However, the predicted change in males dips lower for the CT genotype compared with the CC and TT genotypes. In general, the percentage change in males is much smaller than in females, regardless of genotype. □

*Logistic regression*

As described above, the generalized linear model of Equation (2.9) can also be applied to a binary trait. In this setting, $g()$ is commonly defined as the logit function, reducing Equation (2.9) to the logistic regression model. As described in Section 2.1.1, for a random variable from a Bernoulli trial, we have $E[\mathbf{y}] = Pr(\mathbf{y} = \mathbf{1}_n) = \boldsymbol{\pi}$, where $\mathbf{1}_n$ is an $n \times 1$ vector of 1's, $\boldsymbol{\pi} = (\pi_1, \ldots \pi_n)^T$, and $\pi_i$ is the probability that $y_i = 1$ for $i = 1, \ldots, n$. Thus, we can write Equation (2.10) as

$$g(E[\mathbf{y}]) = \text{logit}(\boldsymbol{\pi}) = \mathbf{X}\beta \tag{2.21}$$

Logistic regression models provide a setting for modeling dichotomous outcomes based on multiple categorical or continuous predictors. The general form of a univariate logistic model in scalar form is given by

$$\text{logit}(\pi_i) = \beta_0 + \beta_1 x_i \tag{2.22}$$

where $\pi_i = Pr(y_i = 1|x_i)$ and $\text{logit}(\pi_i) = \ln\left[\pi_i/(1 - \pi_i)\right]$. For example, suppose $y$ is an indicator for disease status. The $\beta$ parameter is then interpreted as the effect of one unit increase in $x$ on the log odds of disease. If $x$ is again a binary variable for the presence of a variant allele, then $\beta$ is the log odds of disease for individuals with this variant versus those that are homozygous wildtype. In this case, we have $OR = \exp[\beta_1]$. Again additional variables can be added to this model to account for potential confounding and effect modification. Estimation of the parameters is achieved using maximum likelihood methods. Tests of these parameters can again be carried out based on the Wald statistic and LRT. The reader is referred to Hosmer and Lemeshow (2000) and Agresti (2002) for a complete discussion of logistic regression models and associated methods. Finally, fitting a logistic model is also straightforward using the `glm()` function and specifying `family=binomial` in R.

Multivariable regression approaches have many advantages in population-based genetic investigations. First, they provide a natural framework for incorporating continuous predictor variables. In the context of contingency table analysis, continuous variables can be discretized to include in analysis, though this requires *a priori* knowledge about the correct approach for doing this. Another advantage of multivariable models is that they provide a natural setting for inclusion of multiple independent variables. This allows for consideration of many predictors of disease as well as providing a means for assessing the potential confounding or mediating role of additional clinical and demographic factors.

There are some important limitations, however, to keep in mind. First, as a general rule of thumb, limiting our model to include at most one predictor for every five to ten observations for a quantitative trait, or events (given by the minimum of the numbers of successes and failures) in the case of a binary trait, is preferable. In addition, the model parameters may not be identifiable if the predictor variables are highly correlated with one another. Depending on the number of SNPs and their relationship to one another, inclusion of all SNPs within a single model may not be tenable. A first-stage analysis that involves fitting a separate multivariable model for each SNP under investigation, adjusted for confounders and other predictors of the disease trait, may be necessary. Interpreting the results from fitting a large number of such models or testing multiple predictors in a single model needs to include consideration of the number of tests performed. In Chapter 4, several methods for adjusting for multiple testing are described that can be applied to this analysis.

## 2.3 Analytic challenges

The defining characteristics of data arising from genetic association studies render many traditional statistical methods less applicable. This section provides a brief overview of the analytic challenges inherent in characterizing associations among genotypes and measures of disease progression. This serves as a motivation for the remaining chapters of this text, which describe methods aimed at addressing each of these challenges.

### 2.3.1 Multiplicity and high dimensionality

Some of the greatest challenges in the analysis of population association studies arise from what is often called the *curse of dimensionality*, a term originally coined by the applied mathematician Richard Bellman. Investigations involving a large number of genetic markers that are potentially informative have two analytical challenges: (1) inflation of error rates, which arises due to multiple testing; and (2) the complex, generally uncharacterized relationships among the genetic markers under consideration. The former is often referred to as the *multiplicity* problem, and combined these represent the challenge of *high-dimensional* data.

*Error inflation*

In statistical hypothesis testing, we only want to reject the null hypothesis if we are reasonably certain that the alternative hypothesis is in fact true. The probability of making a mistake in this case (that is, the probability of rejecting the null in favor of an alternative when in fact the null is true) is called the *type-1 error rate*. In testing a hypothesis, investigators traditionally aim to control the type-1 error at a level $\alpha$. That is, we want to be certain that the type-1 error rate is less than or equal to $\alpha$. Formally, for a given null hypothesis denoted $H_0$, we have

$$\text{type-1 error rate } = Pr(\text{reject } H_0 \mid H_0 \text{ is true}) \leq \alpha \qquad (2.23)$$

Recall that the *p-value* for a given hypothesis is determined based on a sample of data and is defined as the probability of observing something as extreme or more extreme, given that the null hypothesis is true. If the $p$-value is less than $\alpha$ (typically 0.05), then we reject the null hypothesis in favor of the alternative.

Now suppose we are interested in testing $K$ null hypotheses, given by $H_{0k}$, for $k = 1, \ldots, K$. The *family-wise error under the complete null* (FWEC) is defined as the probability of rejecting one or more of these null hypotheses given that they are all true. If each test is independent and controlled at level $\alpha$, then we have

$$
\begin{aligned}
FWEC &= Pr(\text{reject at least one } H_{0k} \mid H_{0k} \text{ is true for all } k) \\
&= 1 - Pr(\text{reject no } H_{0k} \mid H_{0k} \text{ is true for all } k) \qquad (2.24) \\
&\leq 1 - (1 - \alpha)^K
\end{aligned}
$$

If we are testing only a single hypothesis so that $K = 1$ and we let $\alpha = 0.05$, then Equation (2.24) reduces to $FWEC \leq 0.05$. On the other hand, if we consider two independent tests ($K = 2$), then we are only certain that the FWEC is less than or equal to $1 - 0.95^2 = 0.0975$. This upper bound increases rapidly such that for $K = 10$ independent tests, we have $FWEC \leq 0.401$. In other words, if we conduct ten independent tests, each at a level $\alpha$, then the chance of making a type-1 error is as much as 40.1%. This phenomenon is referred to as *inflation* of the type-1 error rate and is a grave concern in the context of analyzing the associations between a large number of SNPs and a trait. Chapter 4 describes methods for controlling the family-wise error rate as well as the *false discovery rate*, which has been described as a potential alternative error measure in this context. Methods for both independent and potentially correlated tests are presented.

*Unknown model of association*

An additional challenge that arises in the context of a large number of SNPs is that the SNPs are likely to interact with one another in a manner that is not yet well characterized. That is, the model of association between the genotypes and the trait is unknown. The term *interaction* has taken on multiple meanings in the scientific literature. For example, in many biological settings, the term refers to belonging to the same pathway to disease. In a statistical sense, interaction refers to a specific mathematical relationship in which the effect of one polymorphism is modified by the presence of another polymorphism or level of a covariate, as described in Section 2.1.2. Further discussion on the array of definitions associated with the term *interaction* can be found in Cordell (2002), Ahlbom and Alfredsson (2005) and Berrington de Gonzalez and Cox (2007), with an R coding example given in Kallberg *et al.* (2006).

Suppose for example we have a sample of $n$ individuals and $M$ measured SNPs denoted for individual $i$ by $x_{i1}, \ldots, x_{iM}$. For simplicity, suppose each of the $x$ variables are binary indicators for the presence of at least one copy of the corresponding minor allele. Further suppose we are interested in using these variables to predict a quantitative trait $y$. If the SNPs together have an additive effect on the trait, then a typical linear regression model for this setting could be constructed as

$$
\begin{aligned}
y_i &= \beta_0 + \beta_1 x_{i1} + \beta_2 x_{i2} + \ldots + \beta_M x_{iM} + \epsilon_i \\
&= \beta_0 + \sum_{j=1}^{M} \beta_j x_{ij} + \epsilon_i
\end{aligned}
\tag{2.25}
$$

where $i = 1, \ldots, n$ and the $\epsilon_i$'s are assumed to be independent and identically distributed. In this case, a null hypothesis of interest might be that there is no effect on $y$ of having a variant allele at site $j$. Formally, we write this null hypothesis as $H_0 : \beta_j = 0$.

Alternatively, we may believe that the presence of some polymorphisms can potentially alter the effects of other polymorphisms, so that we have interaction in a statistical sense. In this case, an appropriate multivariable regression model is given by

$$
y_i = \beta_0 + \sum_{j=1}^{M} \beta_j x_{ij} + \sum_{k,l(k \neq l)} \gamma_{kl} x_{ik} x_{il} + \epsilon_i
\tag{2.26}
$$

Here the $\gamma_{kl}$ are called interaction effects and are interpreted as the increase (or decrease) in the outcome $y$ that occurs when a variant allele is present at both sites $k$ and $l$; that is, beyond the sum of the effects of a polymorphism at either site alone. Consider for example the simple model in which $M = 2$ and Equation (2.26) reduces to

$$
y_i = \beta_0 + \beta_1 x_{i1} + \beta_2 x_{i2} + \gamma_{12} x_{i1} x_{i2} + \epsilon_i
\tag{2.27}
$$

In this case, the effect of having a variant at site 1 ($x_{i1} = 1$) when $x_{i2} = 0$ is $\beta_1$ and the effect of having a variant at site 2 ($x_{i2} = 1$) when $x_{i1} = 0$ is $\beta_2$; however, the effect of having a variant allele at both sites is $\beta_1 + \beta_2 + \gamma_{12}$, which is more than the sum of the two individual effects.

For the general model given in Equation (2.26), the number of interaction parameters is equal to

$$
\binom{M}{2} = \frac{M!}{(M-2)!2!}
\tag{2.28}
$$

where $M!$ (read "M-factorial") $= M(M-1)(M-2) \ldots 1$. If for example $M = 10$, the number of $\gamma_{kl}$'s is equal to $(10 * 9)/2 = 45$. Clearly, incorporation of three- and four-way interactions would make this model unwieldy. In general, the need for considering potential higher-order interactions is not known. Even for these relatively simple models, unique estimates of the parameters $\beta$ and $\gamma$ may not exist. There are two reasons this can occur. First, if the number of individuals $n$ is less than the number of parameters, given by $(1 + M)$ in Equation (2.25) and $(1 + M + \binom{M}{2})$ in Equation (2.26), then the parameters are not identifiable. In addition, if *collinearity* exists between the SNPs, then unique estimates will also not be obtainable. Collinearity refers to a strong correlation, and in Chapter 3 we will see that this is captured by a high degree of linkage disequilibrium between SNP loci. The problem of estimability can be seen to result from a singular (or close to singular) design matrix and is described further in more advanced linear models texts such as Christensen (2002).

## 2.3.2 Missing and unobservable data considerations

Additional analytic challenges arise from the presence of both missing and unobservable data. In Example 1.1, we encountered missing genotype data on one SNP, and while we ignored it in our analysis, the missingness mechanism may be informative. For example, it may have been more difficult to genotype the rare allele, and thus our frequency estimates would be incorrect. Throughout this text, we use the term *unobservable data* to refer to information that is not observed due to technological limitations. In the genetics setting, the primary example of unobservable data is the alignment of alleles on a single homologous chromosome, commonly referred to as the *haplotypic phase*. This is distinguished from the usual missing data, which are comprised of data that were intended to be collected.

Haplotypic phase is described in Section 1.2 and discussed in more detail in Chapter 5. Briefly, two alleles that are on the same homologous chromosome are said to be *in cis*, while alleles on opposite homologs are said to be *in trans*. The particular combination of alleles on a single strand is called the *haplotype*. Consider for example two biallelic SNPs with alleles $A$, $a$ and $B$, $b$, respectively. As described in Section 1.2, the four possible haplotypes corresponding to these two SNPs are $\{(AB), (Ab), (aB), (ab)\}$. The number of possible haplotypes increases rapidly as the number of SNPs increases. Specifically, if there are $k$ biallelic SNPs, then there are $2^k$ possible haplotypes so that three SNPs yields eight haplotypes and four SNPs results in 32 possible haplotypes. An example of the set of possible haplotype pairs across two biallelic SNPs is given in Figure 2.4. Here the two alleles at the first SNP locus are denoted $A$ and $a$, while the two alleles at the second SNP locus are given by $B$ and $b$. Recall that in a diploid population each individual carries exactly two haplotypes. In the case of $k = 2$ SNPs, there are four haplotypes and $\binom{4}{2} + 4 = 10$ possible combinations of two haplotypes, since an individual can have two copies of the same haplotype.

In general, in population-based association studies, the allelic phase is not observed but is potentially informative. That is, in many instances an individual's pair of haplotypes is not known with complete certainty; however,

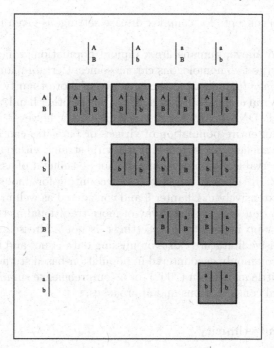

**Fig. 2.4.** Possible haplotype pairs corresponding to two SNPs

information on phase may be relevant for predicting the trait of interest. Consider for example an individual whose genotypes are $Aa$ and $Bb$ at each of two SNPs. Recall that this individual is said to be heterozygous at each of these SNPs. The two possible pairs of haplotypes, also referred to as diplotypes, are $(AB, ab)$ and $(Ab, aB)$, as we saw in Figure 1.2. Notably, the true underlying diplotype is not known, but the probabilities that it is $(AB, ab)$ or $(Ab, aB)$ can be estimated. Methods for arriving at these estimates are described in detail in Chapter 5.

Now consider the setting in which the true disease-causing variant tends to occur on the same segment of DNA as the $AB$ haplotype. That is, we say the "disease allele" is in high linkage disequilibrium (LD) with the $A$ and $B$ alleles. The concept of LD is described in detail in Section 3.1. In this case, an individual with diplotype $(AB, ab)$ would likely exhibit the disease trait, while an individual with the haplotype pair $(Ab, aB)$ would probably not exhibit the disease phenotype. Note that the presence of the disease allele does not always lead to the disease phenotype. The extent to which this phenomenon occurs is termed *penetrance*. Further discussion of penetrance and the related concept of *phenocopies* (individuals who exhibit the disease phenotype but do not carry the allele under consideration), and particularly the estimation of genotype relative risk under departures from full penetrance and no phenocopies, is given in Ziegler and Koenig (2007). A succinct introduction to these and related genetics concepts for complex disease settings is given in Lander and Schork (1994).

As discussed above, humans are a diploid population, which means that each person carries two homologous chromosomes. Certain plants and animal species are *polyploidy*, indicating the presence of more than two homologous chromosomes. Viral and parasitic organisms, on the other hand, are often comprised of a single DNA or RNA strand. Interestingly, however, as seen in the HIV example, an entire population of viruses or parasites can infect a single individual. The allelic phase across multiple viral strains within a single host is similarly unobservable. Unobservable data can be thought of as a special case of missing data. In this text, methods for accounting for unobservable phase are described extensively in Chapter 5 and considered as well in Section 7.1.2. Missing information that arises by way of more traditional mechanisms, such as loss to follow-up in longitudinal settings, is not addressed specifically in this text. Extensive literature exists on missing data issues and methods since missingness is commonly encountered in population-based studies. The reader is referred to Little and Rubin (2002) for a comprehensive summary of appropriate statistical considerations and approaches.

### 2.3.3 Race and ethnicity

Race and ethnicity are amorphous categories that are typically self-reported in population-based investigations. While these categories remain vague, consideration of race and ethnicity is potentially informative in genetic association studies, and understanding the potential implications of population substructure is vital to making valid conclusions from population-based investigations. The terms *population substructure* and *population stratification* are used to refer to the phenomenon in which a population consists of subgroups within which there is random mating but between which there is little mixing or gene flow. An *admixed population* is defined formally as a population in which mating occurs between subgroups with different allelic distributions. However, more commonly population admixture is used loosely to indicate a population in which multiple subgroups are present.

The reasons for considering race and ethnicity in our analysis are multifaceted and described more fully in the following chapters. First, the allelic distributions can vary widely across racial and ethnic groups. In fact, in some instances, the major and minor alleles differ across geographic regions. As a result, ignoring admixture can result in erroneous conclusions about the presence of LD, as described in Section 3.1. In addition, the phenotypic characteristics under investigation can differ dramatically across racial or ethnic groups. For example, it has been reported that Black non-Hispanics tend to have a better lipid profile than White non-Hispanics. As a result of these two phenomena, we commonly encounter the problem of confounding by race or ethnicity and must adjust our analysis appropriately. The concept of confounding by race and ethnicity is discussed in greater detail in Chapter 3.

Another notable difference that is observed across racial and ethnic groups is the length of LD blocks. These are described in more detail in Section 3.1 and are defined as regions of DNA that tend to be conserved from past generations. As a result, the SNP loci that are reported to *tag* these blocks will differ across these groups. Haplotype-based analyses thus generally require a stratified analysis where strata are defined according to race and ethnicity groupings. Finally, race and ethnicity are often thought to capture information on unobserved environmental or demographic factors, such as diet or family history. In that sense, it may be informative in an analysis as any measured clinical factor tends to be. In summary, inclusion of race and ethnicity as main effects in a regression model or as modifiers of the genotype effect on the trait of interest may be the most appropriate model of association.

### 2.3.4 Genetic models and models of association

In this text, a distinction is made between genetic models and models of association. The term *model of association* is used to refer to a mathematical formula relating genotype variables to the trait of interest. For example, in Section 2.3.1, additive and multiplicative models of association were described by Equations (2.25) and (2.26), respectively. Models of association can be much more elaborate, including both deterministic and stochastic elements and complex structure. For example, in Chapter 6 we will see that classification and regression trees impose a model of association that includes a series of conditional associations.

In addition to the model of association, careful consideration needs to be given to what is termed the *genetic model*. The genetic model describes the biological interaction between alleles on homologous chromosomes. Examples of common genetic models are dominant, recessive and additive models. Consider the simple case in which the possible alleles are $A$ and $T$ at a given SNP locus where $A$ is the wildtype or most common allele and $T$ is the variant or minor allele. Further suppose we are interested in a quantitative trait, $y$. An additive model implies that if the effect of having a single copy of the $T$ allele is to increase $y$ by an amount equal to $\beta$, then having two copies of the $T$ allele will increase $y$ by $2\beta$. If we let $I(x_{i,k} = T)$ be an indicator for whether the allele on the $k$th homolog ($k = 1, 2$) is equal to $T$ for individual $i$, then an additive genetic model for this SNP locus is written explicitly as

$$y_i = \alpha + \beta \left[ I(x_{i,1} = T) + I(x_{i,2} = T) \right] + \epsilon_i \tag{2.29}$$

A dominant genetic model, on the other hand, assumes that having one or more copies of the $T$ allele will result in a $\beta$ increase in the quantitative trait $y$ and is given by

$$y_i = \alpha + \beta I(x_{i,1} = T \text{ or } x_{i,2} = T) + \epsilon_i \tag{2.30}$$

Finally, a recessive model assumes that both homologs must contain the rare allele in order for the effect to be present. Formally, this model is written

$$y_i = \alpha + \beta I(x_{i,1} = T) * I(x_{i,2} = T) + \epsilon_i \qquad (2.31)$$

It is important to note that any model of association can be coupled with an underlying genetic model. For example, consider the dominant genetic model of Equation (2.30). An additive model of association with a dominant genetic model across all $M$ loci is written

$$y_i = \alpha + \sum_{j=1}^{M} \beta_j I(x_{ij,1} = T \text{ or } x_{ij,2} = T) + \epsilon_i \qquad (2.32)$$

where $x_{ij,k}$ is the observed base on the $k$th homolog at the $j$th locus for individual $i$. Thus we see that the way in which we define our model inputs, for example the predictor variables in a regression model, reflects an implicit assumption about both the underlying genetic model and model of association. In many instances, both of these models are not known. Understanding the modeling assumptions is critical since specification of an incorrect genetic model can potentially cloud our interpretation of the results from fitting a model of association.

## Problems

**2.1.** Define and contrast each of the following terms: (1) confounding, (2) effect mediation, (3) effect modification, (4) causal pathway, (5) interaction and (6) conditional association.

**2.2.** Based on the FAMuSS data, determine whether any of the four SNPs within the akt2 gene are associated with percentage change in non-dominant arm muscle strength as measured by NDRM.CH. Perform your analysis unadjusted and then adjusting for Race, Gender and Age. State clearly how you code all variables and justify this approach. (No adjustment for multiple testing is needed for this exercise.)

**2.3.** Use the FAMuSS data and the lm() function in R to test for a linear trend in the number of variant alleles at actn3_r577x on percentage change in the non-dominant arm muscle strength as measured by NDRM.CH. Perform this test within and across gender strata. Hint: Begin by coding actn3_r577x as a numeric variable.

**2.4.** Based on the FAMuSS data, determine the odds of having a baseline body mass index, measured by pre.BMI, greater than 25 for individuals who are homozygous variant compared with those that are homozygous wildtype at the resistin_c180g SNP.

**2.5.** Write a one-way ANOVA model with three groups using matrix notation. Assume the total sample size is $n$ and the numbers of individuals in each group are $n_1$, $n_2$ and $n_3$, respectively. Let the quantitative response be given by $\mathbf{y}$.

**2.6.** Write an additive model of association that assumes a recessive genetic model. Be sure to define all terms explicitly.

**2.7.** Suppose we conduct five independent hypothesis tests for association, each at a level $\alpha = 0.05$. Determine the upper limit for our resulting type-1 error rate. What is this limit if instead we conducted 50 independent $\alpha$-level tests?

# 3

# Genetic Data Concepts and Tests

This chapter provides an overview of two important concepts relevant to population-based association studies: linkage disequilibrium (LD) and Hardy-Weinberg equilibrium (HWE). Both of these concepts are based on the genetic component of our data and do not involve the trait that we aim to associate with this genetic information. Instead, consideration of LD and HWE is an important step of data processing that typically precedes consideration of association, and thus a discussion is included in this text. Both LD and HWE are measures of allelic association—the difference between them is that LD is a measure of allelic association between two sites on the genome, while HWE is a measure of allelic association between two homologous chromosomes at a single site.

## 3.1 Linkage disequilibrium (LD)

In candidate gene studies, the hypothesis of interest is whether the gene is involved in the causal pathway to disease. In this case, the particular SNP loci within the gene that are chosen may not be functional; that is, they may not directly cause the disease. However, these sites are likely to be *associated* with disease because they are in what is commonly referred to as *linkage disequilibrium* (LD) with the functional variant. Linkage disequilibrium is defined as an association in the alleles present at each of two sites on a genome. Notably, the concept of LD differs from the term *linkage* used in the context of linkage analysis. Linkage analysis is an approach that aims to identify the location on a chromosome of a specific gene. Linkage analysis is typically applied in the context of family-based studies (i.e., studies involving related individuals) and draws on information from genes with known location. Linkage analysis is based on the phenomenon that the farther apart genes are, the more likely that a recombination event has occurred between them. Recombination in the genetic sense is defined as the joining of two broken DNA strands, one from the maternal side and one from the paternal side, and occurs as parental

A.S. Foulkes, *Applied Statistical Genetics with R: For Population-based Association* 65
*Studies*, Use R, DOI: 10.1007/978-0-387-89554-3_3,
© Springer Science+Business Media LLC 2009

chromosomes are passed to an offspring. Further details on recombination in the context of meiosis are provided in Section 1.3.1.

Fine mapping studies are often performed to further investigate regions of a chromosome that were identified through linkage analyses. These studies again aim to identify the location of a specific candidate gene, in this case with greater precision. Both fine mapping and linkage analysis differ from candidate-gene association studies in that a fundamental goal of the former two is to *map* the chromosomal *location* of a disease-causing gene. The specific allele(s) contributing to the disease phenotype are less relevant. Candidate-gene and candidate-polymorphism studies instead aim to characterize the association between allelic variation at a given location and the disease phenotype. Notably, methods that simultaneously assess linkage and association have been described. For a complete discussion of linkage, the reader is referred to Thomas (2004).

The following section defines two closely related measures of LD: $D'$ and $r^2$. These measures are also related to Pearson's $\chi^2$-statistic, typically generated for a test of association between two categorical variables, as described in Chapter 2. Several additional measures of LD have been described, each with its own set of advantages and disadvantages. Additional discussion of these measures can be found in Devlin and Risch (1995), Chapter 8 of Thomas (2004) and Chapter 9 of Ziegler and Koenig (2007).

### 3.1.1 Measures of LD: $D'$ and $r^2$

To begin, consider the distribution of alleles for $n$ individuals across two sites. Let us first assume that the two sites are independent of one another. That is, the presence of an allele at one site does not influence the particular allele observed at the second site. Further, suppose $A$ and $a$ are the possible alleles at Site 1, $B$ and $b$ are the alleles at Site 2, and $p_A$, $p_a$, $p_B$ and $p_b$ denote the population frequencies for $A$, $a$, $B$ and $b$, respectively. Since each individual carries two homologous chromosomes, there are in total $N = 2n$ homologs across the $n$ individuals in our sample. The expected distribution of alleles under independence between Sites 1 and 2 is given in Table 3.1. Here the number in each cell of this $2 \times 2$ table is the corresponding haplotype count and is denoted $n_{ij}$ for $i, j = 1, 2$. For example, $n_{11}$ is the expected number of homologs with allele $A$ at Site 1 and allele $B$ at Site 2 in this population. That is, it is the expected number of homologs with the $AB$ haplotype. Recall from Section 2.1.1 that, under independence, the frequency of $AB$ is simply the product of the frequency of $A$ and the frequency of $B$. Mathematically, we write $p_{AB} = p_A p_B$, where $p_{AB}$ is the joint probability of $A$ and $B$ occurring together. Therefore, the expected number of $AB$ homologs is given by $n_{11} = N p_{AB} = N p_A p_B$.

If the observed data support Table 3.1, then this provides evidence that Sites 1 and 2 are in linkage equilibrium. If, on the other hand, Sites 1 and 2 are in fact associated with one another, then the observed counts will deviate from

**Table 3.1.** Expected allele distributions under independence

|        |   | Site 2 | | |
|--------|---|--------|--------|--------|
|        |   | $B$ | $b$ | |
| Site 1 | $A$ | $n_{11} = Np_Ap_B$ | $n_{12} = Np_Ap_b$ | $n_{1.} = Np_A$ |
|        | $a$ | $n_{21} = Np_ap_B$ | $n_{22} = Np_ap_b$ | $n_{2.} = Np_a$ |
|        |   | $n_{.1} = Np_B$ | $n_{.2} = Np_b$ | $N = 2n$ |

**Table 3.2.** Observed allele distributions under LD

|        |   | Site 2 | | |
|--------|---|--------|--------|--------|
|        |   | $B$ | $b$ | |
| Site 1 | $A$ | $n_{11} = N(p_Ap_B + D)$ | $n_{12} = N(p_Ap_b - D)$ | $n_{1.}$ |
|        | $a$ | $n_{21} = N(p_ap_B - D)$ | $n_{22} = N(p_ap_b + D)$ | $n_{2.}$ |
|        |   | $n_{.1}$ | $n_{.2}$ | $N = 2n$ |

the numbers in Table 3.1. The amount of such deviation is represented by the scalar $D$ in Table 3.2. With the introduction of $D$, this table represents the more general setting, in which independence across sites is not assumed. Note that if $D = 0$, then Table 3.2 reduces to Table 3.1. Intuitively, the value of the scalar $D$ captures information on the magnitude of the departure from linkage equilibrium. For example, if $D$ is relatively large, then the counts $n_{11}$ and $n_{22}$ in Table 3.2 will be greater than is expected under independence, while $n_{12}$ and $n_{21}$ will be smaller than expected, suggesting a large departure from LD. On the other hand, if the absolute value of $D$ is close to 0, then the observed counts in Table 3.2 will be close to the expected numbers under independence, indicating little or no departure from LD. The two measures described in this chapter, $D'$ (pronounced D-prime) and $r^2$ (pronounced R-squared), are both functions of the scalar $D$.

We can express $D$ in terms of the joint probability of $A$ and $B$ and the product of the individual allele probabilities as follows:

$$D = p_{AB} - p_Ap_B \qquad (3.1)$$

In practice, we estimate $D$ by plugging in the corresponding estimates of the marginal and joint probabilities. It is straightforward to show that $\widehat{p}_A = n_{1.}/N$ and $\widehat{p}_B = n_{.1}/N$ are the estimates of $p_A$ and $p_B$, respectively. Estimation of $p_{AB}$ is not as straightforward, however, since the number of homologs in our sample with the $A$ and $B$ alleles is not observed. That is, the number of copies of the $AB$ haplotype is uncertain, resulting from the fact that we have

**Table 3.3.** Genotype counts for two biallelic loci

|        |     | Site 2 | | |
|--------|-----|--------|--------|--------|
|        |     | $BB$   | $Bb$   | bb     |
| Site 1 | $AA$ | $n_{11}$ | $n_{12}$ | $n_{13}$ |
|        | $Aa$ | $n_{21}$ | $n_{22}$ | $n_{23}$ |
|        | $aa$ | $n_{31}$ | $n_{32}$ | $n_{33}$ |

population-based data of unrelated individuals, as described in Sections 1.3.1 and 2.3.2.

To estimate $p_{AB}$ in this setting, we first write the likelihood of our parameters, $\theta = (p_{AB}, p_{Ab}, p_{aB}, p_{ab})$, conditional on the observed data. Recall from Section 2.1.1 that the likelihood is written as the joint probability of the observed data, in this case the cell counts. Thus, we have

$$
\begin{aligned}
\log L(\theta | n_{11}, \ldots, n_{33}) \propto{} & (2n_{11} + n_{12} + n_{21}) \log p_{AB} \\
& + (2n_{13} + n_{12} + n_{23}) \log p_{Ab} + (2n_{31} + n_{21} + n_{32}) \log p_{aB} \quad (3.2) \\
& + (2n_{33} + n_{32} + n_{23}) \log p_{ab} + n_{22} \log(p_{AB}p_{ab} + p_{Ab}p_{aB})
\end{aligned}
$$

where now $n_{ij}$ are the genotype counts, for $i, j = 1, 2, 3$, as shown in Table 3.3. This likelihood can be rewritten as a function of $p_{AB}$, $p_A$ and $p_B$ by using the relationships $p_{Ab} = p_A - p_{AB}$, $p_{aB} = p_B - p_{AB}$ and $p_{ab} = 1 - p_A - p_B - p_{AB}$. Finally, since there is no closed-form solution for the maximum likelihood estimate of $p_{AB}$, a numerical algorithm, such as Newton-Raphson, can be applied. In general, numerical algorithms are cumbersome when more than two sites are considered, and alternative approaches, such as an expectation-maximization (EM) type algorithm, are applied for haplotype frequency estimation, as described in Chapter 5.

Since cell counts cannot be negative, the value of $D$ is constrained in a way that depends on $p_A$, $p_a$, $p_B$ and $p_b$. For this reason, a rescaled value of $D$, given by $D'$, is often used as a measure of LD. Formally,

$$
D' = \frac{|D|}{D_{\max}} \quad (3.3)
$$

where $D_{\max}$ represents the upper bound on $D$ and is given by:

$$
D_{\max} = \begin{cases} \min\,(p_A p_b, p_a p_B) & D > 0 \\ \min\,(p_A p_B, p_a p_b) & D < 0 \end{cases} \quad (3.4)
$$

Note that $0 \leq D' \leq 1$. Values of $D'$ that are close to 1 are assumed to indicate high levels of LD, while values close to 0 suggest low LD. Importantly, formal testing of the hypothesis that LD is equal to 0 requires consideration of the fact that haplotype frequencies used in the calculation of $D$ are estimated. Inference procedures must account for the additional variability resulting from this estimation. Further discussion of the challenges inherent in testing for linkage disequilibrium in the context of phase ambiguity is provided in Schaid (2004). The following two examples illustrate how to calculate an estimate of $D'$ between pairs of SNPs for data collected on unrelated individuals using R.

*Example 3.1 (Measuring LD using $D'$).* In this example, we use the R package genetics. Loading this package requires several additional R packages that are not included in the typical R installation, including MASS, combinat, gdata, gtools and mvtnorm. Additional details on installing new packages can be found in the appendix. To load the necessary R packages, we use the library() function as follows:

```
> library(genetics)
```

Suppose we aim to calculate LD as measured by $D'$ for two SNPs within the gene alpha-actinin 3 (actn3) based on data from the FAMuSS study. The reader is referred to Section 1.3.3 for additional background on these data. We begin by attaching these data so that we can call associated variables:

```
> attach(fms)
```

The next step is to create objects of class genotype for each of our SNP variables. Recall from Table 1.1 that the first ten observations for SNPs r577x and rs540874 have the following form:

```
> actn3_r577x[1:10]

 [1] CC CT CT CT CC CT TT CT CT CC
Levels:  CC CT TT

> actn3_rs540874[1:10]

 [1] GG GA GA GA GG GA AA GA GA GG
Levels:  AA GA GG
```

To create genotype objects, we use the corresponding genotype() function as follows:

```
> Actn3Snp1 <- genotype(actn3_r577x,sep="")
> Actn3Snp2 <- genotype(actn3_rs540874,sep="")
```

Note that we specify sep="" since in our data the alleles at a given SNP site are not separated. For example, the genotype for the first person at the first SNP is CC. The default character assumed to delineate alleles within a site is /. We can see by printing the first ten observations that the data for the first SNP are now represented as follows:

```
> Actn3Snp1[1:10]
```

```
[1] "C/C" "C/T" "C/T" "C/T" "C/C" "C/T" "T/T" "C/T" "C/T" "C/C"
Alleles: C T
```

with the corresponding class attributes given by

```
> class(Actn3Snp1)
```

```
[1] "genotype" "factor"
```

The data are now appropriately formatted for calculating $D'$. We do this using
the LD() function of the genetics package:

```
> LD(Actn3Snp1,Actn3Snp2)$"D'"
```

```
[1] 0.8858385
```

This result, $D' = 0.89$, suggests that there is a high degree of LD between
SNPs labeled r577x and rs540874 within the actn3 gene.

If instead we consider two SNPs across two different genes, we expect $D'$ to
be relatively small. For example, consider the SNP labeled rs1801132 within
the estrogen receptor 1 gene (esr1). $D'$ between this SNP locus and the first
SNP in actn3 yields a much lower value, suggesting that there is not strong
LD between these SNPs. This is illustrated with the following R code:

```
> Esr1Snp1 <- genotype(esr1_rs1801132,sep="")
> LD(Actn3Snp1,Esr1Snp1)$"D'"
```

```
[1] 0.1122922                                                          □
```

*Example 3.2 (Measuring pairwise LD for a group of SNPs).* Now suppose we
are interested in calculating pairwise LD for a group of SNPs within the actn3
gene. In this case, we can use the same R function, LD(). First we read in two
additional SNPs as genotype variables and create a corresponding dataframe:

```
> Actn3Snp3 <- genotype(actn3_rs1815739,sep="")
> Actn3Snp4 <- genotype(actn3_1671064,sep="")
> Actn3AllSnps <- data.frame(Actn3Snp1,Actn3Snp2,Actn3Snp3,Actn3Snp4)
```

A matrix of pairwise LD measures is then given by the upper triangular
elements of the $D'$ matrix:

```
> LD(Actn3AllSnps)$"D'"
```

```
          Actn3Snp1 Actn3Snp2 Actn3Snp3 Actn3Snp4
Actn3Snp1        NA 0.8858385 0.9266828 0.8932708
Actn3Snp2        NA        NA 0.9737162 0.9556019
Actn3Snp3        NA       NA         NA 0.9575870
Actn3Snp4        NA        NA        NA        NA
```

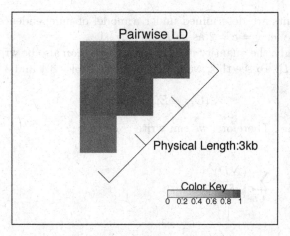

**Fig. 3.1.** Map of pairwise LD

As expected since these SNPs are all within the same gene, there appears to be a high level of LD for all pairs. That is, in all cases, the presence of a given allele at one SNP locus tends to be associated with the presence of a specific allele at another SNP locus. This result can also be illustrated using the `LDheatmap()` function within the `LDheatmap` package, as follows:

```
> install.packages("LDheatmap")
> library(LDheatmap)
> LDheatmap(Actn3AllSnps, LDmeasure="D'")
```

The resulting plot is given in Figure 3.1, where the shading represents the degree of association.                                               □


Another intuitively appealing measure of LD is the quantity $r^2$. This measure is based on Pearson's $\chi^2$-statistic for the test of no association between the rows and columns of an $r \times c$ contingency table such as the one given in Table 3.2. Specifically, $r^2$ is defined as

$$r^2 = \chi_1^2/N \tag{3.5}$$

Recall that as described in Section 2.2.1, Pearson's $\chi^2$-test statistic is given by

$$\chi_1^2 = \sum_{i,j} \frac{(O_{ij} - E_{ij})^2}{E_{ij}} \tag{3.6}$$

where $i = 1, 2, \ldots, r$, $j = 1, 2, \ldots, c$, and $O_{ij}$ and $E_{ij}$ are respectively the observed and expected cell counts for the $i, j$ cell of an $r \times c$ table. The expected counts are determined under a model of independence, and in the LD setting we let $r = c = 2$ as in Table 3.2.

Interestingly, the quantity $r^2$ in Equation (3.5) can also be written in terms of the scalar $D$. To see this, we first note from Tables 3.1 and 3.2 that

$$(O_{ij} - E_{ij})^2 = (ND)^2 \tag{3.7}$$

for all $i, j$ pairs. Therefore, we can write

$$\chi_1^2 = \sum_{i,j} \frac{(ND)^2}{E_{ij}}$$

$$= (ND)^2 \left( \frac{1}{N p_A p_B} + \frac{1}{N p_A p_b} + \frac{1}{N p_a p_B} + \frac{1}{N p_a p_b} \right) \tag{3.8}$$

$$= ND^2 \left( \frac{p_a p_b + p_a p_B + p_A p_b + p_A p_B}{p_A p_B p_a p_b} \right)$$

$$= \frac{ND^2}{p_A p_B p_a p_b}$$

The final equality holds since $p_a + p_A = 1$ and $p_b + p_B = 1$. Thus we have

$$r^2 = \chi_1^2 / N = \frac{D^2}{p_A p_B p_a p_b} \tag{3.9}$$

Notably, the difference between $D'$ and $r^2$ rests in the type of adjustment made to the scalar $D$. In both cases, this adjustment involves the marginal allele frequencies since the value of $D$ will depend on these. Investigators tend to prefer $r^2$ due to its straightforward relationship to the usual $\chi^2$-test for a contingency table analysis. Readers are cautioned, however, against comparing $Nr^2$ to a $\chi_1^2$-distribution in the analysis of population-based data. Recall that in population-based investigations, haplotypes are not observed and so the cell counts of the corresponding contingency table are not known. As a result, an estimation procedure, such as the EM algorithm, must be used, and in turn this introduces additional variability into our measure. In addition, Pearson's $\chi^2$-test assumes independent observations, which may be violated in our sample in the absence of Hardy-Weinberg equilibrium (HWE) (described below) since the contingency table includes two observations per person. This challenge is highlighted in Sasieni (1997). The following example illustrates estimation of LD based on $r^2$.

*Example 3.3 (Measuring LD based on $r^2$ and the $\chi^2$-statistic).* Suppose that we are again interested in measuring LD between the SNPs `r577x` and `rs540874` within the `actn3` gene based on the FAMuSS data. Calculation of $r^2$ can be achieved using the same `LD()` function described in Examples 3.1 and 3.2. Again the SNP variables first need to be transformed into genotype objects:

```
> attach(fms)
> library(genetics)
> Actn3Snp1 <- genotype(actn3_r577x,sep="")
> Actn3Snp2 <- genotype(actn3_rs540874,sep="")
> LD(Actn3Snp1,Actn3Snp2)$"R^2"
```

```
[1] 0.6179236
```

Similar to Example 3.1, based on this measure there is strong evidence to suggest LD between these two SNP loci.

This result is based on the $n = 725$ individuals with complete data on both SNPs. Based on this, we see that the corresponding $\chi^2$-statistic is equal to $0.6179236 * 725 * 2 = 896$. The `LD()` function also returns this statistic and a corresponding "*p*-value", as seen below with the complete function output:

```
> LD(Actn3Snp1,Actn3Snp2)
```

```
Pairwise LD
-----------
                 D          D'       Corr
Estimates: 0.1945726 0.8858385 0.7860811

            X^2 P-value    N*
LD Test: 895.9891       0 725
```

```
(*note: N in this output is represented by n in the textbook notation.)
```

The reader is cautioned however against interpreting this statistic as a formal test of LD (as suggested by the corresponding package documentation) for two reasons. First, it is generated based on two observations per person. That is, the total of the cell counts is equal to $N = 2 * n$, where $n$ is our sample size. Second, cell counts are estimated, which introduces additional variability. Thus, the usual $\chi^2$-test for association is not valid in this setting.    □

$D'$ and $r^2$ are both *measures* of linkage disequilibrium between loci, estimating the amount of association between sites. While the information captured by these quantities can be highly useful, conclusions must be drawn with caution. These quantities reflect estimates of association and interpreting them as formal statistical tests has a few limitations. First, as noted in Example 3.3, the $\chi^2$-statistic that is generated using the `LD()` function in R is based on data from a $2 \times 2$ table that includes correlated data—in this case, two observations for each individual. Second, an additional layer of estimation is performed in the process of calculating these measures since the haplotypic phase and thus the cell counts are not observable. Accounting for the error associated with this estimation is crucial for testing association between loci. Thus this statistic cannot be interpreted in the usual way described for contingency table analysis.

### 3.1.2 LD blocks and SNP tagging

In the previous section, we considered measures of pairwise LD between two alleles. More generally, interest may lie in determining whether a group of alleles are in LD. One intuitively appealing measure of LD across a region comprised of multiple SNPs is simply the average of all pairwise measures $D'$. For example, suppose $D'_{ij}$ is a measure of LD between loci $i$ and $j$ for $i, j \in \mathcal{L}$, where $\mathcal{L}$ is a set of loci within a region of interest. Then one measure of LD for this region is given by

$$\bar{D}' = \frac{1}{n_\mathcal{L}} \sum_{i,j \in \mathcal{L}} D'_{ij} \qquad (3.10)$$

where $n_\mathcal{L}$ is the number of ways of choosing two loci from the set $\mathcal{L}$ and the summation is over all such pairs. This calculation is straightforward, as the following example shows.

*Example 3.4 (Determining average LD across multiple SNPs).* Returning to the FAMuSS data and the `actn` gene, we saw in Example 3.2 that calculation of $D'$ for all pairs of SNPs within this gene is straightforward using the LD() function

```
> LDMat <- LD(Actn3AllSnps)$"D'"
```

where `actn3.allSNPs` is a dataframe with each element a genotype object representing a SNP locus within `actn`. Calculating the average LD is achieved using the `mean()` function as in the following code. Remember to specify that missing values should be removed by including `na.rm=T`.

```
> mean(LDMat,na.rm=T)
```

```
[1] 0.9321162
```

More precise estimates of LD are also tenable through fine mapping studies. Ultimately, through characterizing regions of high LD, the human genome can be divided into *LD blocks*. These blocks are separated by *hotspots*, regions in which recombination events are more likely to occur. For further discussion of LD blocks and recombination hotspots, see Balding (2006). An illustration of two LD blocks separated by a recombination hotspot is given in Figure 3.2. In general, alleles tend to be more correlated within LD blocks than across LD

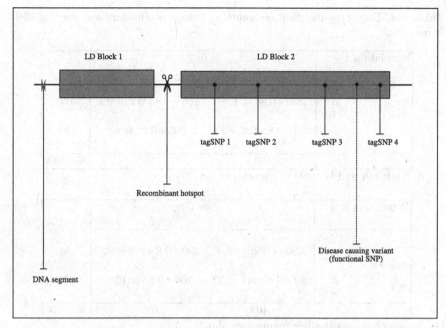

**Fig. 3.2.** Illustration of LD blocks and associated tag SNPs

blocks. Once regions of high LD are identified, investigators aim to determine the smallest subset of SNPs that characterizes the variability in this region, a process referred to as *SNP tagging*. Here the goal is to reduce redundancies in the genetic data. Consider a pair of SNPs that are in perfect LD so that $D'$ is equal to 1. Genotyping both SNPs in this case is unnecessary since the relationship between the two is deterministic. That is, by the definition of LD, knowledge of the genotype of one SNP completely defines the genotype of the second, and so there is no reason to sequence both loci. Well-defined tag SNPs will capture a substantial majority of the genetic variability within an LD block.

The tag SNPs corresponding to one LD block are illustrated in Figure 3.2. As shown in this figure, tag SNPs are correlated with the true disease-causing variant but are not typically functional themselves. Notably, LD blocks differ substantially across race and ethnicity groups and in particular tend to be shorter in Black non-Hispanics than Whites and Hispanics. As a result, a set of tag SNPs may capture information on the true disease-causing variant in one racial group and not another. Consideration of this phenomenon and application of appropriate analytic methods is crucial in the analysis of population-based association studies that include multiple racial and ethnic groups. In the following section, we discuss the impact of population substructure on measures of LD. Importantly, the differences in the SNPs that

**Table 3.4.** Haplotype distribution assuming linkage equilibrium and varying allele frequencies

| Population 1 | | Site 2 | | |
| --- | --- | --- | --- | --- |
| | | $B$ | $b$ | |
| Site 1 | $A$ | $200 * 0.8^2 = 128$ | $200 * 0.8 * 0.2 = 32$ | 160 |
| | $a$ | $200 * 0.8 * 0.2 = 32$ | $200 * 0.2^2 = 8$ | 40 |
| | | 160 | 40 | $N = 200$ |

(a) Assuming A and B allele frequencies of 0.8

| Population 2 | | Site 2 | | |
| --- | --- | --- | --- | --- |
| | | $B$ | $b$ | |
| Site 1 | $A$ | $200 * 0.2^2 = 8$ | $200 * 0.8 * 0.2 = 32$ | 40 |
| | $a$ | $200 * 0.8 * 0.2 = 32$ | $200 * 0.8^2 = 128$ | 160 |
| | | 40 | 160 | $N = 200$ |

(b) Assuming A and B allele frequencies of 0.2

capture information on LD block across racial and ethnic groups may lead investigators to the collection of different variables across these subpopulations, in which case a stratified analysis will be necessary.

### 3.1.3 LD and population stratification

As described in Section 2.3.3, population stratification refers to the presence of multiple subgroups between which there is minimal mating or gene transfer. Ignoring underlying population stratification in our sample can lead to erroneous conclusions about the presence of LD between two SNPs. This is illustrated in the following example.

*Example 3.5 (Population substructure and LD).* Consider two SNPs with dominant allele frequencies $p_A = p_B = 0.8$ in one population and $q_A = q_B = 0.2$ in a second population. Suppose that in both populations these two SNPs are not associated; that is, there is no LD between them. Under such a model of no association, the observed counts will be similar to the counts given in Tables 3.4(a) and (b) for the first and second populations, respectively, assuming a sample size of $n = 100$ individuals ($N = 2 * 100$ haplotypes) for each group. If data from these two populations are combined into a single $2 \times 2$ contingency table, then the observed counts will be given by Table 3.5. The expected counts are calculated in R using the `chisq.test()` function and involves first creating a matrix of observed counts:

**Table 3.5.** Apparent LD in the presence of population stratification

| Populations 1 and 2 | | Site 2 | | |
| --- | --- | --- | --- | --- |
| | | $B$ | $b$ | |
| Site 1 | $A$ | $128 + 8 = 136$ | $32 + 32 = 64$ | 200 |
| | $a$ | $32 + 32 = 64$ | $8 + 128 = 136$ | 200 |
| | | 200 | 200 | $N = 400$ |

```
> ObsCount <- matrix(c(136,64,64,136),2)
> ObsCount

     [,1] [,2]
[1,]  136   64
[2,]   64  136

> ExpCount <- chisq.test(ObsCount)$expected
> ExpCount

     [,1] [,2]
[1,]  100  100
[2,]  100  100
```

Taking the absolute difference between the observed and expected counts and dividing by $N$ yields $D = 36/400 = 0.09$. Furthermore, based on the observed cell counts of the combined populations, we have $p_A = p_a = p_B = p_b = 0.5$ and so $D_{max} = 0.25$ and $D' = 0.09/0.25 = 0.36$. Thus, combining data across the two populations and not accounting for the resulting substructure in our analysis leads to the incorrect conclusion that there is mild LD between the alleles at Sites 1 and 2. □

The effect of population admixture on LD is a well-described concept in biostatistics and epidemiology texts and is commonly referred to as Simpson's paradox; see, for example, Pagano and Gauvreau (2001). Simpson's paradox occurs in the presence of a confounding variable that is not appropriately accounted for in the analysis. In this setting, "population" is confounding the relationship between Sites 1 and 2, and by ignoring this factor we see an association that does not exist in either population on its own. Further discussion of confounding, how to adjust for it appropriately in analysis, and other related concepts is provided in Chapter 2. Notably, population substructure is not always observed, and thus straightforward application of statistical methods for confounding may not be tenable.

## 3.2 Hardy-Weinberg equilibrium (HWE)

Another important concept in population-based genetic association studies is Hardy-Weinberg equilibrium (HWE). While LD refers to allelic association across sites on a single homolog, HWE denotes independence of alleles at a single site between two homologous chromosomes. Consider for example the simple case of one biallelic SNP with genotypes $AA$, $Aa$ and $aa$. HWE implies that the probability of an allele occurring on one homolog does not depend on which allele is present on the second homolog. Formally, independence in this setting is equivalent to stating, for example, that the joint probability of $A$ and $a$, given by $p_{Aa}$, is equal to the product of the individual allele probabilities, $p_A$ and $p_a$. Formally, independence implies $p_{AA} = p_A^2$, $p_{Aa} = p_A p_a$ and $p_{aa} = p_a^2$, where $p_A$ and $p_a = 1 - p_A$ are the population frequencies of alleles $A$ and $a$, respectively. The following section introduces two approaches to testing for a departure from HWE.

### 3.2.1 Pearson's $\chi^2$-test and Fisher's exact test

Tests of HWE include Pearson's $\chi^2$-test and Fisher's exact test. The $\chi^2$-test is computationally advantageous but relies on asymptotic theory. Thus, when more than 20% of the expected counts are less than five, Fisher's exact test is preferable. Consider the $2 \times 2$ table of genotypes at a single locus given in Table 3.6. Here $n_{11}$ and $n_{22}$ are the number of individuals with genotypes $AA$ and $aa$, respectively, and these counts are observed. Notably, the genotypes $Aa$ and $aA$ are indistinguishable in population-based investigations, and thus we only observe the sum $n_{12}^* = n_{21} + n_{12}$ and not the individual cell counts, $n_{21}$ and $n_{12}$.

The expected counts corresponding to these three observed counts, $n_{11}$, $n_{12}^*$, $n_{22}$, are given respectively by $E_{11} = N p_A^2$, $E_{12} = 2N p_A (1 - p_A)$ and $E_{22} = N(1 - p_A)^2$, where $p_A$ is the probability of $A$ and is estimated based on the observed allele count. That is, we let $p_A = (2n_{11} + n_{12}^*)/(2N)$. The $\chi^2$-test statistic is then constructed in the usual way, as described in Section 2.2.1,

$$\chi^2 = \sum_{(i,j) \in \mathcal{C}} \frac{(O_{ij} - E_{ij})^2}{E_{ij}} \sim \chi_1^2 \qquad (3.11)$$

where now the summation is over the set $\mathcal{C}$ of three observed cells. This statistic is compared with the appropriate quantile of a $\chi_1^2$-distribution to determine whether to reject the null hypothesis of HWE. We still have a single degree of freedom since knowledge of the count in any one of the three cells in $\mathcal{C}$ fully determines the counts in the remaining cells, conditional on the marginal totals. For example, if we know $n_{12}^*$, then we can determine $n_{11}$ and $n_{22}$. To see this, note that we have $n_{11} + n_{12} = n_{1.}$ and $n_{11} + n_{21} = n_{.1}$. Therefore, $2n_{11} + n_{12}^* = n_{1.} + n_{.1}$ or equivalently $n_{11} = (n_{1.} + n_{.1} - n_{12}^*)/2$.

**Table 3.6.** Genotype counts for two homologous chromosomes

| | | Homolog 2 | | |
| | | $A$ | $a$ | |
|---|---|---|---|---|
| Homolog 1 | $A$ | $n_{11}$ | $n_{12}$ | $n_{1.}$ |
| | $a$ | $n_{21}$ | $n_{22}$ | $n_{2.}$ |
| | | $n_{.1}$ | $n_{.2}$ | $n$ |

That is, knowledge of $n_{12}^*$ and the margin totals, $n_{1.}$ and $n_{.1}$, is sufficient to determine $n_{11}$. A similar expression can be derived for $n_{22}$.

A statistically significant test of HWE suggests that the SNP under investigation is in Hardy-Weinberg *disequilibrium* (HWD). The term HWD is often used synonymously with the phrase *non-random mating* since HWD can arise from self-selecting mates. The relationship between HWE and population substructure is discussed further in Section 3.2.2. An example of applying Pearson's $\chi^2$-test to test for HWE is given below.

*Example 3.6 (Testing for HWE using Pearson's $\chi^2$-test).* Suppose we are interested in testing for HWE for the SNP labeled `AKT1.C0756A` in the HGDP data. To do this, we first need to calculate the observed and expected genotype counts. This can be done as follows:

```
> attach(hgdp)
> Akt1Snp1 <- AKT1.C0756A
> ObsCount <- table(Akt1Snp1)
> Nobs <- sum(ObsCount)
> ObsCount

Akt1Snp1
 AA  CA  CC
 48 291 724

> FreqC <- (2 * ObsCount[3] + ObsCount[2])/(2*Nobs)
> ExpCount <- c(Nobs*(1-FreqC)^2, 2*Nobs*FreqC*(1-FreqC),Nobs*FreqC^2)
> ExpCount

[1]  35.22319 316.55362 711.22319
```

In this example, vectors of observed and expected counts are for the genotypes $AA$, $CA$ and $CC$, respectively. The $\chi^2$-statistic is calculated using the formula of Equation (3.11) and the following R code:

```
> ChiSqStat <- sum((ObsCount - ExpCount)^2/ExpCount)
> ChiSqStat

[1] 6.926975
```

This statistic has a $\chi^2$-distribution with a single degree of freedom. The quantile corresponding to $1 - \alpha$, where $\alpha = 0.05$, is given by

```
> qchisq(1-0.05,df=1)
```

```
[1] 3.841459
```

Since $6.93 > 3.84$, based on this sample we would reject the null hypothesis of HWE at this SNP locus and conclude instead that the alleles on the two homologous chromosomes are associated with one another. Alternatively, the HWE.chisq() function in the genetics package can be used to calculate this statistic. In this case, we first use the genotype() function to create a genotype object as follows:

```
> library(genetics)
> Akt1Snp1 <- genotype(AKT1.C0756A, sep="")
> HWE.chisq(Akt1Snp1)
```

```
        Pearson's Chi-squared test with simulated p-value (based on
        10000 replicates)
```

```
data:  tab
X-squared = 6.927, df = NA, p-value = 0.007199
```

Note that the same $\chi^2$-statistic is returned with a corresponding $p = 0.0072$. Again, based on this result, we would reject the null hypothesis of HWE and conclude that there appears to be non-random mating. In Section 3.2.2, we return to this example with a further analysis that considers the ethnic diversity of this sample.                                                       □

The $p$-value from Fisher's exact test is based on summing the exact probabilities of seeing the observed count data or something more extreme in the direction of the alternative hypothesis. Fisher showed that the exact probability from a contingency table such as Table 3.6 is given by

$$p_A = \frac{\binom{n_{1.}}{n_{11}}\binom{n_{2.}}{n_{21}}}{\binom{N}{n_{.1}}} = \frac{n_{1.}!\,n_{2.}!\,n_{.1}!\,n_{.2}!}{N!\,n_{11}!\,n_{12}!\,n_{21}!\,n_{22}!} \tag{3.12}$$

It was further shown by Emigh (1980) for the genetics setting that if we let $n_1 = 2 * n_{11} + n_{12}^*$, we can write this probability as

$$p_A = \frac{\binom{n}{n_{11},n_{12}^*,n_{22}}}{\binom{2n}{n_1}} 2^{n_{12}^*} \tag{3.13}$$

In the following example, we illustrate calculation of this exact probability and the exact $p$-value for a test of HWE.

*Example 3.7 (Testing for HWE using Fisher's exact test).* Now suppose we are interested in testing for a departure from HWE for the same SNP of Example 3.6 but only within the Maya population. This subgroup consists of $N = 25$ individuals, and the observed and expected genotype counts are calculated as follows:

```
> attach(hgdp)
> Akt1Snp1Maya <- AKT1.C0756A[Population=="Maya"]
> ObsCount <- table(Akt1Snp1Maya)
> ObsCount

Akt1Snp1Maya
AA CA CC
 1  6 18

> Nobs <- sum(ObsCount)
> FreqC <- (2 * ObsCount[3] + ObsCount[2])/(2*Nobs)
> ExpCount <- c(Nobs*(1-FreqC)^2, 2*Nobs*FreqC*(1-FreqC),Nobs*FreqC^2)
> ExpCount

[1]  0.64  6.72 17.64
```

Since the expected count for the first cell is less than 5, using Fisher's exact test to test for HWE is most appropriate. An exact probability of seeing the observed counts, as described in Equation (3.13), is given by `FisherP1` in the following code example:

```
> n11 <- ObsCount[3]
> n12 <- ObsCount[2]
> n22 <- ObsCount[1]
> n1 <- 2*n11+n12
> Num <- 2^n12 * factorial(Nobs)/prod(factorial(ObsCount))
> Denom <- factorial(2*Nobs) / (factorial(n1)*factorial(2*Nobs-n1))
> FisherP1 <- Num/Denom
> FisherP1

[1] 0.4011216
```

Fisher's exact $p$-value is given by summing over all probabilities of seeing something as extreme as or *more* extreme than the observed data. Thus, to arrive at this $p$-value, we need to perform the calculation above for the more extreme situations as well, given by $n_{11} = 19$, $n_{11} = 20$ and $n_{11} = 21$, with $n_{12}$ and $n_{22}$ adjusted accordingly.

The function `HWE.exact()` from the `genetics` package can also be used to calculate this exact $p$-value, as illustrated below:

```
> library(genetics)
> Akt1Snp1Maya <- genotype(AKT1.C0756A[Population=="Maya"], sep="")
> HWE.exact(Akt1Snp1Maya)
```

Exact Test for Hardy-Weinberg Equilibrium

```
data:  Akt1Snp1Maya
N11 = 18, N12 = 6, N22 = 1, N1 = 42, N2 = 8, p-value = 0.4843
```

Based on this output, we see that the exact $p$-value is 0.4843 and we are unable to reject the null hypothesis that there is a departure from HWE in this population.                                                                  □

## 3.2.2 HWE and population substructure

The concepts of HWE and population substructure are closely linked. In this section, we highlight some important connections that will help guide analysis strategies discussed in later chapters. Specifically, we illustrate that: (1) HWE implies constant allele frequencies over generations; (2) HWE is violated in the presence of population admixture; and (3) HWE is violated in the presence of population stratification.

We begin by demonstrating that, under the HWE assumption, allele frequencies remain constant over generations. Consider for example the genotype of a parent at a single biallelic locus. Assuming HWE, we know that the joint probability of a pair of alleles is the product of the two individual-level probabilities. This is also referred to as statistical independence, as described in Section 2.1.1, and implies that the probabilities of genotypes $AA$, $Aa$ and $aa$ are given by $p_{AA} = p_A^2$, $p_{Aa} = 2p_A(1 - p_A)$ and $p_{aa} = (1 - p_A)^2$, respectively. We also know that the probability that an offspring inherits the $A$ allele from a parent with genotype $AA$ is 1, from a parent with genotype $Aa$ it is $1/2$ and from a parent with genotype $aa$ it is 0. Formally, if $Pr(A|X)$ is the conditional probability of an offspring having the $A$ allele given that the parental genotype is $X$, then we have

$$
\begin{aligned}
Pr(A|AA) &= 1 \\
Pr(A|Aa) &= 1/2 \\
Pr(A|aa) &= 0
\end{aligned}
\tag{3.14}
$$

Thus, the joint probabilities of the parental genotype and inheritance of the $A$ allele are given by

$$
\begin{aligned}
Pr(A, AA) &= Pr(A|AA)Pr(AA) = p_A^2 \\
Pr(A, Aa) &= Pr(A|Aa)Pr(Aa) = p_A(1 - p_A) \\
Pr(A, aa) &= Pr(A|aa)Pr(aa) = 0
\end{aligned}
\tag{3.15}
$$

using the identities stated in Section 2.1.1. The frequency of $A$ in this new generation is equal to the sum of the probabilities in Equation (3.15). That

**Table 3.7.** Example of the effect of population admixture on HWE

| Population 1 | Population 2 | | |
| --- | --- | --- | --- |
| | $A$ | $a$ | |
| $A$ | $np_A q_A = 64$ | $np_A(1 - q_A) = 96$ | $np_A = 160$ |
| $a$ | $n(1 - p_A)q_A = 16$ | $n(1 - p_A)(1 - q_A) = 24$ | $n(1 - p_A) = 40$ |
| | $nq_A = 80$ | $n(1 - q_A) = 120$ | $n = 200$ |

is, $Pr(A) = Pr(A, AA) + Pr(A, Aa) + Pr(A, aa) = p_A$ and is the same as the parent generation. This result implies that allele frequencies will tend to remain constant over time under HWE.

In the presence of population admixture, on the other hand, a departure from the HWE assumption will likely be detected. Recall from Section 2.3.3 that *population admixture* refers to the setting in which mating occurs between two populations for which the allele frequencies differ. For example, suppose $Pr(A) = p_A$ in one population and $Pr(A) = q_A$ in a second population, where $p_A \neq q_A$. Now suppose two individuals, one from each population, reproduce. The probability that the offspring has the $AA$ genotype is given by $p_A q_A$, the probability of the $Aa$ genotype is $p_A(1 - q_A) + q_A(1 - p_A)$ and the probability of the $aa$ genotype is $(1 - p_A)(1 - q_A)$. Now consider a sample of $n = 200$ individuals from an admixed population where $Pr(A) = p_A = 0.8$ in one subpopulation and $Pr(A) = q_A = 0.4$ in the second sub-population for a given site. In this setting, the observed cell count data will look similar to those given in Table 3.7.

Recall that the observed data will consist of $n_{11} = 64$, the sum of the two off-diagonal cells, given by $n_{12}^* = 96 + 16 = 112$, and $n_{22} = 24$. Under the HWE assumption, the corresponding expected counts are given by the quantities $E_{11} = Np_0^2 = 72$, $E_{12} = 2Np_0(1 - p_0) = 96$ and $E_{22} = N(1 - p_0)^2 = 32$, where $p_0$ is the estimated $A$ allele frequency, $p_0 = (2 * 64 + 96 + 16)/(2 * 200) = 0.6$. Thus, the $\chi^2$ statistic corresponding to a test of HWE, for this example, is given by

$$\chi_1^2 = \left( \frac{(72 - 64)^2}{72} + \frac{(96 - 112)^2}{96} + \frac{(32 - 24)^2}{32} \right) = 5.56 \qquad (3.16)$$

Comparing this result to a $\chi^2$-distribution with 1 degree of freedom, we would reject the null hypothesis of HWE. In other words, inbreeding of two populations with differing allele frequencies (i.e., population admixture) results in an apparent departure from HWE.

A similar result is observed in the context of population stratification. Recall that *population stratification* is the combination of populations in which breeding occurs within but not between subpopulations. Consider for example

**Table 3.8.** Genotype distributions for varying allele frequencies

| Population 1 | | Homolog 2 | | |
|---|---|---|---|---|
| | | $A$ | $a$ | |
| Homolog 1 | $A$ | $200 * 0.8^2 = 128$ | $200 * 0.8 * 0.2 = 32$ | 160 |
| | $a$ | $200 * 0.8 * 0.2 = 32$ | $200 * 0.2^2 = 8$ | 40 |
| | | 160 | 40 | $n = 200$ |

(a) Assuming $Pr(A) = 0.8$

| Population 2 | | Homolog 2 | | |
|---|---|---|---|---|
| | | $A$ | $a$ | |
| Homolog 1 | $A$ | $200 * 0.4^2 = 32$ | $200 * 0.4 * 0.6 = 48$ | 80 |
| | $a$ | $200 * 0.4 * 0.6 = 48$ | $200 * 0.6^2 = 72$ | 120 |
| | | 80 | 120 | $n = 200$ |

(b) Assuming $Pr(A) = 0.4$

the genotype counts for two populations, given by Tables 3.8(a) and 3.8(b), in which the probabilities of the $A$ allele are $Pr(A) = 0.8$ and $Pr(A) = 0.4$, respectively. Within each population, we have HWE since the observed cell counts are as expected under random mating. If, however, we consider the data on the two populations combined, as given in Table 3.9, it can be shown that the HWE assumption is violated. This is commonly referred to as the Wahlund effect, after the Swedish geneticist who first documented the concept (Wahlund, 1928). To see this phenomenon, we can use the `chisq.test()` function in R, assuming for simplicity of presentation that the cell counts are fully observed:

```
> ObsDat <- matrix(c(160,80,80,80),2)
> chisq.test(ObsDat,correct=FALSE)

        Pearson's Chi-squared test

data:  ObsDat
X-squared = 11.1111, df = 1, p-value = 0.0008581
```

Here we specify `correct=FALSE`, indicating that Yates' continuity correction is not necessary since the expected cell counts are all greater than 5. Based on this result, we would reject the null of HWE and conclude there is disequilibrium in this stratified population. Notably, this is very similar to the result that we saw in the context of linkage disequilibrium, described in Section 3.1.3. The following example illustrates the effect of geographic origin on HWE for one SNP within the HGDP data.

**Table 3.9.** HWD in the presence of population stratification

| Populations 1 and 2 | | Homolog 2 | | |
|---|---|---|---|---|
| | | $A$ | $a$ | |
| Homolog 1 | $A$ | $128 + 32 = 160$ | $32 + 48 = 80$ | 240 |
| | $a$ | $32 + 48 = 80$ | $8 + 72 = 80$ | 160 |
| | | 240 | 160 | $n = 400$ |

*Example 3.8 (HWE and geographic origin).* Returning to the HGDP data described in Section 1.3.3, we notice that there are individuals from multiple geographic regions. The distribution of the $n = 1064$ individuals across these regions is given by

```
> attach(hgdp)
> table(Geographic.area)
```

```
Geographic.area
 Central Africa Central America         China           Israel
           119              50           184              148
         Japan     New Guinea Northern Africa Northern Europe
            31              17            30               16
      Pakistan         Russia   South Africa   South America
           200              67             8               58
Southeast Asia Southern Europe
            11             125
```

Tests of HWE within each region are calculated using the `tapply()` and `HWE.chisq()` functions. In the following code, we print out the results for two regions for the `AKT1.C0756A` SNP:

```
> library(genetics)
> Akt1Snp1 <- genotype(AKT1.C0756A, sep="")
> HWEGeoArea <- tapply(Akt1Snp1,INDEX=Geographic.area,HWE.chisq)
> HWEGeoArea$"Central Africa"

        Pearson's Chi-squared test with simulated p-value (based on
        10000 replicates)

data:  tab
X-squared = 0.2322, df = NA, p-value = 0.6589

> HWEGeoArea$"South America"

        Pearson's Chi-squared test with simulated p-value (based on
        10000 replicates)

data:  tab
X-squared = 27.2386, df = NA, p-value = 9.999e-05
```

A review of the output for all regions reveals that population admixture or stratification is likely to be present within the observed South American and Russian samples. In all other geographic areas, we cannot detect a deviation from the HWE assumption based on a level $\alpha = 0.05$ $\chi^2$-test unadjusted for multiple comparisons.                                                                    □

In practice, a test of HWE is used to assess whether either population admixture or stratification is present. A test of HWE is also used to identify genotyping errors, as described in Section 3.3.2 below. While admixture and stratification represent two different phenomena—the former describes in-breeding while the latter implies the presence of multiple subpopulations in which there is no inbreeding—the manifestation of both is a violation of the HWE assumption. The analytic implications are therefore similar and the terms are often used interchangeably. Throughout the remainder of the text, the phrase *population substructure* is used generically to encompass both population admixture and population stratification.

## 3.3 Quality control and preprocessing

In the remaining chapters of this textbook, we focus on methods for identifying and characterizing association among multiple SNPs and a trait. Prior to this analysis, careful consideration of potential errors in the data, through the application of quality control measures, is essential. The need for this is particularly relevant to genome-wide association (GWA) studies in which a vast array of data are measured. Here we give a brief discussion of SNP chips (Section 3.3.1) and methods for identifying and accounting for genotyping errors (Section 3.3.2), population substructure (Sections 3.3.3 and 3.3.5) and relatedness (Section 3.3.4).

### 3.3.1 SNP chips

The introduction of SNP chip technology, coupled with the success of the International Haplotype Map (HapMap) Project, has led to an explosion of new clinical research studies involving whole or partial genome-wide scans, commonly referred to as genome-wide association studies (GWAS). High-throughput SNP genotyping platforms, including Affymetrix and Illumina chips, provide for simultaneous genotyping of $500,000$ to one million SNPs. While the human genome consists of approximately $3 \times 10^9$ bases, variability in the genome is captured by a subset of SNPs due to well-defined LD blocks. In fact, evidence suggests that whole genome-wide arrays of approximately one million SNPs are sufficient to characterize human genetic variability across a population.

Detailed information on chip technology can be found in Kennedy *et al.* (2003) and Affymetrix (2006). Notably, the output of a SNP array is a *probe quartet* for each possible pair of alleles for all SNPs on the chip. This probe quartet consists of an ensemble of continuous measures: two intensities based on perfect match probes, one for each of the pair of alleles, and two intensities corresponding to mismatch probes. For example, for the two alleles $A$ and $T$, at each site we have measures of probe intensity corresponding to: (1) a perfect match for $A$; (2) a perfect match for $T$; (3) a mismatch for $A$; and (4) a mismatch for $T$. Application of a classification-type algorithm yields a *genotype call*, for example $AA$, $AT$ or $TT$, based on these data. One such algorithm that exhibits relatively strong performance is based on a robustly fitted linear model and an application of Mahalanobis distance (RLMM), as described in Rabbee and Speed (2006). This approach, as well as appropriate data normalization, can be applied in R using the `Classify()` and `normalize_Rawfiles()` functions within the `RLMM` package. While the concordance between genotype calls and the true underlying genotype is remarkably high using a sophisticated algorithm such as that of Rabbee and Speed (2006), some error tends to be introduced at this stage. Methods for detecting remaining genotyping errors are discussed in Section 3.3.2 below.

The focus of this text is on candidate gene investigations; however, the statistical methods described herein are equally relevant and applicable to GWAS. Several software tools have been developed that apply these methods while accounting for the computational demands of whole genome-wide investigations. Included in these is, most notably, the free and open-source package PLINK, which can be downloaded for Linux, MS-DOS and Mac OS at the PLINK webpage: http://pngu.mgh.harvard.edu/~purcell/plink/. PLINK includes applications of many of the methods described in this text and can be used in combination with R to take advantage of the array of statistical tools in R and the advanced data-handling ability of PLINK using `Rserve`. Additional information and details can be found in Purcell *et al.* (2007). The R packages `SNPassoc`, `GENAbel` and `snpMatrix` are also designed specifically to handle genome-wide association data. The `snpMatrix` package is a component of the BioConductor open-source and open development software project for the analysis of genomic data, and additional details can be found in Clayton and Leung (2007). While specific applications of the functions provided within these packages are not illustrated in this text, the concepts behind many of these functions are described, including tests of association such as the Cochran-Armitage trend test and Pearson's $\chi^2$-test, measures of linkage disequilibrium such as $D'$ and $r^2$, and summaries of population substructure, including principal component analysis.

While GWAS have gained a great deal of popularity in recent years, the need remains for well-designed studies that investigate candidate genes, their interactions, and the potential modifying and confounding roles of clinical and demographic data. The findings from GWAS will aid in these investigations by offering novel hypotheses about the pathways to disease. These studies will also provide new, better characterized candidate genes requiring further study in a more traditional epidemiological context.

### 3.3.2 Genotyping errors

A *genotyping error* is defined as a deviation between the true underlying genotype and the genotype that is observed through the application of a sequencing approach. These errors occur with varying degrees of frequency across the
different technological platforms and arise for a variety of different reasons.
For a complete discussion of causes and frequencies of genotyping errors, see
Ziegler and Koenig (2007). The most common statistical approach to identifying genotyping errors in population-based studies of unrelated individuals
is testing for a departure from HWE at each of the SNPs under investigation (Hosking *et al.*, 2004). This proceeds in the same manner as described
in Section 3.2.1 using either an asymptotic $\chi^2$-test or Fisher's exact test for
association. There are a few notable drawbacks to this approach that we describe here, as they are important to keep in mind in making the decision to
remove SNPs from analysis.

First, deviations from HWE within individuals with disease (cases) can
be due to association between genotypes and disease status. In order to account for this possibility, some researchers advocate testing for departures
from HWE only within the control population. This, however, can be mis-
leading as well since removing cases from the analysis may lead to an apparent departure from HWE when indeed the entire population is in HWE.
One resolution to this problem for case–control studies involves application
of a goodness-of-fit test to identify the most probable genetic disease model;
e.g., dominant, recessive, additive or multiplicative (Wittke-Thompson *et al.*,
2005). This approach provides a means of distinguishing between departures
from HWE that are due to the model for disease prevalence and those departures that are indeed a result of other phenomena, such as genotyping
errors.

Another drawback to using tests of HWE exclusively for identifying genotyping errors is that a departure from HWE may in fact be a result of population substructure, as described in Section 3.2.2. Removing these SNPs from
the analysis can therefore lead to a loss of data that are potentially informative regarding underlying structure in our population. In Section 3.3.3 below,
we discuss methods for identifying population substructure that use this information. Thus, it is generally recommended that use of HWE to identify
genotyping errors be coupled with repeat genotyping. Unfortunately, repeat
genotyping is costly and labor intensive and therefore impractical in many
instances.

Finally, multiplicity is also a challenge in this setting. As described in
Section 2.3.1, multiple testing leads to an inflation of the type-1 error rate.
Therefore, if we test for a departure of HWE at each of multiple SNPs, then
the likelihood of incorrectly rejecting the null hypothesis (and concluding
HWD) can be substantial, particularly in the context of GWAS. This is at-
tenuated to some extent by the correlated nature of the tests, arising from LD
among SNPs; however, consideration of multiple testing adjustments is still

warranted. In this setting, a more extreme threshold for significance, such as 0.005 or 0.001, is typically applied. A more extensive discussion of multiple testing procedures is given in Chapter 4.

If a departure from HWE is detected and additional investigation confirms that there is truly a genotyping error, then the entire SNP is typically removed from analysis. That is, these approaches do not provide for identification of specific records (individuals) within which an error exists. Instead, it is assumed that the genotype information is incorrect across all individuals in the sample and the entire corresponding column is removed prior to fitting and interpreting models of association.

### 3.3.3 Identifying population substructure

As discussed in Sections 2.1.2 and 2.3.3, the presence of population substructure can result in spurious associations. There are, broadly speaking, two approaches to handling population substructure in the context of association studies. The first is to stratify the analysis by racial and ethnic groups and in some instances, in particular if the corresponding strata are small, remove outlying individuals prior to testing for association. A second approach is to account for the population substructure in the analysis of association. Both approaches are complicated by the fact that race and ethnicity are not welldefined. That is, while information on a self-declared value of race and ethnicity is often available in population-based investigations, this variable, in a true sense, is unobservable, or what is commonly referred to as *latent*. In this section, we focus on methods for identifying population substructure. Based on the findings from this analysis, we can proceed with (1) a stratified or subset analysis or (2) an adjustment in the analysis of association using a multivariable modeling approach, as described in Section 2.2.3.

Applications of principal components analysis (PCA) and multidimensional scaling (MDS) (also known as principal coordinate analysis) provide visual means of identifying population substructure. The idea behind both approaches is to provide a low-dimensional representation of our data that captures information on the variability between individuals across SNPs. These methods are described in detail in many multivariate texts; see for example Johnson and Wichern (2002). Briefly, the aim of MDS is to fit our data into a lower dimensional space (coordinate system) such that the pairwise distances between individuals are similar to the original distances (in the higher dimensional space.) The aim of PCA, on the other hand, is to identify $k$ ($k < p$) linear combinations of the data, commonly referred to as principal components, that capture overall variability, where $p$ is the number of variables, or SNPs in our setting.

MDS thus begins by defining a measure of *similarity* between all pairs of individuals. For example, for a given individual, we can let each SNP be represented by a 0, 1 or 2, corresponding to the number of variant alleles present at the corresponding site. The *similarity* between two individuals is then defined as the distance between the respective vectors of data. The most commonly applied measure of distance is Euclidean, though alternative measures, such as the Manhattan and binary distances, may be reasonable options. An example of generating a similarity matrix is illustrated in the following example.

*Example 3.9 (Generating a similarity matrix).* In this example we generate distances between all pairs of individuals in the FAMuSS dataset based on all 24 SNPs within the `akt1` gene. To do this, we apply the `dist()` function in R. First we create a vector containing the names of all of the SNPs within this gene as follows:

```
> attach(fms)
> NamesAkt1Snps <- names(fms)[substr(names(fms),1,4)=="akt1"]
> NamesAkt1Snps
```

```
 [1] "akt1_t22932c"        "akt1_g15129a"
 [3] "akt1_g14803t"        "akt1_c10744t_c12886t"
 [5] "akt1_t10726c_t12868c" "akt1_t10598a_t12740a"
 [7] "akt1_c9756a_c11898t"  "akt1_t8407g"
 [9] "akt1_a7699g"         "akt1_c6148t_c8290t"
[11] "akt1_c6024t_c8166t"  "akt1_c5854t_c7996t"
[13] "akt1_c832g_c3359g"   "akt1_g288c"
[15] "akt1_g1780a_g363a"   "akt1_g2347t_g205t"
[17] "akt1_g2375a_g233a"   "akt1_g4362c"
[19] "akt1_c15676t"        "akt1_a15756t"
[21] "akt1_g20703a"        "akt1_g22187a"
[23] "akt1_a22889g"        "akt1_g23477a"
```

The `substr()` function extracts elements of the character string given in its argument. By applying this function to a vector, we extract these elements from each component of the vector and return a vector. In this case, we specify that we want to take the one to four elements of each of the character strings in the vector of names associated with the `fms` data.

The next step is to convert the genotype data from factor variables to numeric variables using the `data.matrix()` function. Note that we additionally assign the missing data a number in the code below:

```
> FMSgeno <- fms[,is.element(names(fms),NamesAkt1Snps)]
> FMSgenoNum <- data.matrix(FMSgeno)
> FMSgenoNum[is.na(FMSgenoNum)] <- 4
```

This results in values of 1, 2, 3 and 4 for each SNP. In all cases, the number 4 corresponds to missing (`NA`) and the number 2 corresponds to the heterozygous genotype, while the numbers 1 and 3 can correspond to either homozygous wildtype or homozygous variant at the corresponding site. Finally, we apply the `dist()` function to compute the distance matrix for the resulting dataframe and print the results for the first five individuals:

```
> DistFmsGeno <- as.matrix(dist(FMSgenoNum))
> DistFmsGeno[1:5,1:5]

           1          2          3          4          5
1 0.000000  4.795832  5.291503  3.741657  3.162278
2 4.795832  0.000000  2.236068  3.872983  3.000000
3 5.291503  2.236068  0.000000  3.162278  3.741657
4 3.741657  3.872983  3.162278  0.000000  2.449490
5 3.162278  3.000000  3.741657  2.449490  0.000000
```

This table tells us that the Euclidean distance between individuals 1 and 2, for example, is 4.796, while the distance between individuals 1 and 5 is 3.162.

□

Several R functions provide for application of MDS, including cmdscale(), which is an application of the classical MDS method, sammon() and isoMDS() in the MASS package, and several functions within the vegan and SensoMineR packages, which offer extensions. For example, sammon() and isoMDS() apply non-metric multidimensional scaling based on an initial rank ordering of pairwise distances. The three functions listed require a similarity matrix as input. An example of applying the classical approach is given below followed by an application of PCA.

*Example 3.10 (Multidimensional scaling (MDS) for identifying population substructure).* Suppose we are interested in determining whether there is any evidence for population substructure in the FAMuSS cohort based on the akt1 SNPs. We can begin with the dataset labeled FMSgenoNum and the corresponding distance matrix, given by DistFmsGeno, that we generated in Example 3.9. We then plot the first and second coordinates from an MDS analysis as follows:

```
> plot(cmdscale(DistFmsGeno),xlab="C1",ylab="C2")
> abline(v=0,lty=2)
> abline(h=4,lty=2)
```

The resulting plot is given in Figure 3.3, which suggests that there may be multiple clusters in the data. Specifically, there appear to be as many as three clusters within the lower left quadrangle of the plot, as well as two clusters that are more clearly delineated in the top left and bottom right quadrangles. Notably, there is substantial missing data in this cohort that may be driving the formation of these clusters. Furthermore, the missingness mechanism appears to be related to race/ethnicity, and thus interpretation of these findings must be approached with caution.    □

*Example 3.11 (Principal components analysis (PCA) for identifying population substructure).* Similar results are found in the application of PCA. We use the following R code to determine the principal components based on the SNP dataset generated in Example 3.9 and to plot the data in the space defined by the first and second principal components:

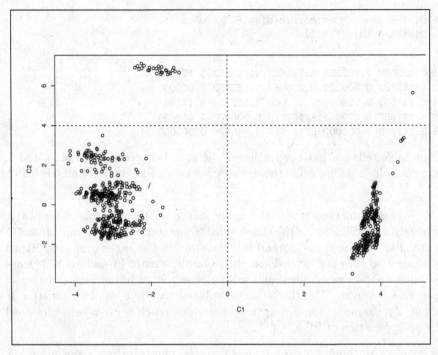

**Fig. 3.3.** Application of MDS for identifying population substructure

```
> PCFMS <- prcomp(FMSgenoNum)
> plot(PCFMS$"x"[,1],PCFMS$"x"[,2],xlab="PC1",ylab="PC2")
```

The resulting plot is illustrated in Figure 3.4. For this example, the plot reveals information identical to what we saw in Figure 3.3.                                      □

### 3.3.4 Relatedness

In the methods described throughout this text, we assume individuals are unrelated. If this assumption is violated, then, as described in Section 1.1.3, accounting for the within-family correlations is imperative for valid inference. Unfortunately, relatedness between individuals can easily arise in large-scale population-based studies but is not always known to exist. Several quality control measures have been described for detecting relationship misspecifications among individuals in a sample. Here we describe one relatively simple and intuitively appealing approach, called the graphical representation of relationship errors (GRR), proposed by Abecasis *et al.* (2001). The idea behind GRR is that pairs of individuals who have the same relationship to one another (e.g., siblings, parent–offspring pairs or unrelated individuals) will share a similar number of *identical-by-state (IBS)* alleles. Alleles that are *IBS* have

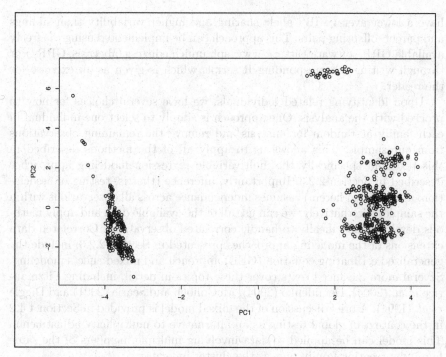

**Fig. 3.4.** Application of PCA for identifying population substructure

the same DNA composition and may or may not derive from the same ances-tor. *Identical-by-descent (IBD)* alleles derive from the same ancestor. Further discussion of these concepts can be found in Ziegler and Koenig (2007) and Thomas (2004).

GRR begins by enumerating the shared IBS alleles across multiple SNPs between all pairs of individuals within a given type of relationship. In population-based investigations, the assumption is that all individuals are unrelated and thus there is only a single relationship type. Suppose we let $x_{ij,k}$ be the number of IBS alleles for individuals $i$ and $j$, $i \neq j$, at SNP $k$, where $k = 1, \ldots, K$. Note that for a biallelic locus $k$, $x_{ij,k}$ takes on the value 0, 1 or 2. The mean and variance of IBS allele sharing for individuals $i$ and $j$ across the $K$ SNPs are then given respectively by

$$\mu_{ij} = \frac{1}{K} \sum_{k=1}^{K} x_{ij,k} \qquad (3.17)$$

$$\sigma_{ij}^2 = \frac{1}{K-1} \sum_{k=1}^{K} (x_{ij,k} - \mu_{ij})^2 \qquad (3.18)$$

A scatterplot of $(\mu_{ij}, \sigma_{ij})$ for all pairs of individuals then reveals potential relationship misspecifications. In general, we expect unrelated individuals to have a lower average IBS allele sharing and higher variability than siblings and parent–offspring pairs. This approach can be implemented using the freely available GRR software (http://www.sph.umich.edu/csg/abecasis/GRR/) or through writing a corresponding R script, which is given as an exercise for the reader.

Upon identifying related individuals, we have several choices for how to proceed with the analysis. One approach is simply to select one individual in each family at random for analysis and remove the remaining observations from our sample. This allows us to apply all of the methods described in this text and specifically the multivariable regression modeling approaches described in Section 2.2.3. Importantly, inference (that is, testing of association, as described herein) assumes independence across all observations within the sample. Alternatively, we can use all of the available data and apply methods developed specifically to handle correlated observations. Correlated data extensions of the modeling approaches presented in Section 2.2.3, include the generalized estimating equation (GEE) approach and mixed effects modeling. Several more advanced texts cover these topics in detail, including Fitzmaurice *et al.* (2004), Demidenko (2004), McCulloch and Searle (2001) and Diggle *et al.* (1994). A brief discussion of the mixed model is provided in Section 4.4.2 in the context of global testing as an alternative to multiplicity adjustments. This model can be applied to data involving multiple members of the same family by treating family unit as the cluster indicator.

### 3.3.5 Accounting for unobservable substructure

In Section 3.3.3 and 3.3.4, we describe the use of MDS, PCA and GRR to identify the presence of population substructure and relatedness in our sample. Based on the results from this analysis, we may choose to stratify our sample for subsequent analysis or simply remove individuals who appear to deviate from well-defined racial/ethnic groupings. This, however, is not always practical since such groupings are generally amorphous, with a broad spectrum of deviations. In addition, by stratifying our analysis in the absence of effect modification, we lose power to detect true underlying associations. An alternative is to treat population heterogeneity and/or cryptic relatedness as a confounder in our analysis. This, however, poses an analytic challenge since both phenomena are generally not observed. Several methods have been proposed in the context of case–control studies to handle this situation, including genomic control (Devlin and Roeder, 1999), structured association (Pritchard *et al.*, 2000), principal components analysis (Price *et al.*, 2006) and the stratification-score approach (Epstein *et al.*, 2007).

Here we describe briefly the most recently proposed stratification-score approach. This approach is intuitively appealing, relies on having access to data on a collection of substructure informative loci and proceeds in two stages. The motivation for the stratification-score approach is that population substructure, represented by the variable $U$, is a potential confounder and therefore estimation of genotype effects without adjusting for $U$ is invalid. However, $U$ is unobservable, so direct adjustment is not tenable. Instead, the stratification-score approach makes use of the fact that we can define strata based on substructure informative loci within which the effect of our genotypes of interest can be estimated correctly. Suppose $Z$ represents the substructure informative loci and $X$ is the genotype under study. We begin in stage 1 by calculating the odds of disease given $Z$, denoted $\theta_Z$, using for example a simple logistic regression model. More sophisticated models can also be applied at this stage. If we assume that the unobservable variable $U$ does not interact in a statistical sense with $X$, then the effect of $X$ on the odds of disease can be estimated within strata with a constant value of $\theta_Z$. Thus, in stage 2, we test for association between $X$ and disease status within each of five strata defined based on quartiles of $\theta_Z$. Information on software availability for the stratification-score approach can be found at http://www.genetics.emory.edu/labs/epstein/software/.

The challenge of unmeasured confounding has received a lot of attention in the statistical literature since it is common in observational (non-randomized) studies. The broad literature on causal inference and instrumental variables techniques may serve as a backbone for further methodological developments to handle unobservable substructure. For an introduction and overview of causal inference, see for example Pearl (2000) and Gelman and Meng (2004).

# Problems

**3.1.** Define and contrast the following terms: (a) linkage disequilibrium (LD), (b) Hardy-Weinberg equilibrium (HWE), (c) population stratification, (d) population admixture and (e) tagSNP.

**3.2.** Determine whether there is a deviation from Hardy-Weinberg equilibrium (HWE) for the akt1_t10726c_t12868c SNP based on the full FAMuSS cohort. Check whether stratifying by the variable Race alters your findings. Interpret the results of your analysis.

**3.3.** Report an overall measure of linkage disequilibrium (LD) for all SNPs within the esr1 gene for the FAMuSS data. Does this measure adequately summarize the pairwise estimates of LD between these SNPs?

**3.4.** Report estimates of pairwise linkage disequilibrium (LD) for all SNPs within the akt1 gene for the HGDP data. Do these estimates tend to vary across Geographic.area? Interpret your findings.

**3.5.** Assess whether there is any evidence for genotyping errors in the akt1 gene for the HGDP data. State clearly how you made this determination.

**3.6.** Determine whether there is any evidence for population substructure in African Americans in the FAMuSS data. Explain how you reached this conclusion.

**3.7.** Write an R script to determine the mean and standard deviation of identical-by-state (IBS) allele sharing between all pairs of observations. Derive an example and plot and interpret the results.

# 4

# Multiple Comparison Procedures

In the previous chapter, several traditional statistical methods were described for testing for an association between a genotype and a trait. These methods generally require comparing a test statistic to its known (or simulated) distribution in order to quantify the probability of seeing what we do or something more extreme under the assumption that the null hypothesis is true. Our decision to accept or reject the null then rests on a comparison of this quantity, called the *p-value*, to a threshold based on a previously determined acceptable level of error. In population-based association studies, we generally aim to test for the presence of associations between the trait and each of multiple genotypes across several SNPs and gene loci. As we saw in Section 2.3 and describe in more detail below, testing multiple hypotheses can result in an inflation of the error rate, which we want to control. Several methods, termed *simultaneous test procedures (STPs)*, have been developed to address this challenge directly. In this chapter, a few such methods for adjusting for multiple testing are described, including single-step and step-down methods (Section 4.2) and resampling-based approaches (Section 4.3). First, some important measures of error are defined (Section 4.1). The advanced reader is referred to the more theoretical coverage of multiple testing procedures and their latest developments in Dudoit and van der Laan (2008), Dudoit *et al.* (2003) and Chapter 16 of Gentleman *et al.* (2005). These discussions include applications to data arising from gene expression studies, human genetic association studies and HIV genetic investigations.

## 4.1 Measures of error

Much of the literature on methods for adjusting for multiple comparisons describes controlling one of two error rates: the family-wise error rate (FWER) and the false discovery rate (FDR). In this section, we define each of these measures and describe their relative interpretations in population association methods. Additional discussions of FWER and FDR can be found in Westfall

A.S. Foulkes, *Applied Statistical Genetics with R: For Population-based Association Studies*, Use R, DOI: 10.1007/978-0-387-89554-3_4,
© Springer Science+Business Media LLC 2009

**Table 4.1.** Type-1 and type-2 errors in hypothesis testing

|       |       | Test |  |
|-------|-------|------|--|
|       |       | Non-significant | Significant |
| Truth | $H_0$ | True Negative | **Type-1 Error** |
|       | $H_A$ | **Type-2 Error** | True Positive |

and Young (1993) and Benjamini and Hochberg (1995), respectively. Methods for error control in the context of multiple testing are described in Sections 4.2 and 4.3.

### 4.1.1 Family-wise error rate

Table 4.1 illustrates the four scenarios for testing hypotheses and the types of errors that can result. The rows in this table correspond to the (unobserved) truth, and the columns correspond to the result of applying our testing procedure. Tests of a single hypothesis can result in two types of errors: (1) concluding that the null hypothesis is false when in fact it is true, called *type-1 error*, and (2) not rejecting the null hypothesis when it is indeed false, called *type-2 error*. Consider for example the null hypothesis that the mean response for genotype 1, given by $\mu_1$, is equal to the mean response for genotype 2, given by $\mu_2$. Formally, this null is written $H_0 : \mu_1 = \mu_2$. Further suppose we conduct a two-sample $t$-test of this null hypothesis, and the resulting test statistic is given by $T$. The first type of error, a type-1 error, occurs when this test statistic is declared significant and in fact the null is true. In our example, this means that $T$ is more extreme than the predefined significance threshold but in fact the two population means are equal. The second type of error, referred to as a type-2 error, occurs when we fail to reject the null hypothesis but it is in fact false. For our example, this would mean that $T$ is less than the significance threshold but the two population means are actually different.

Traditionally, statistical testing relies on control of the type-1 error rate. The *level* of a test, usually denoted by $\alpha$ and set equal to 0.05, is precisely the probability of incorrectly rejecting the null hypothesis. That is, the level of a test is the probability of making a type-1 error. Typically the $p$-value is compared to $\alpha$ in order to make a decision about whether or not to reject the null hypothesis. The *p-value* is defined formally as the probability of observing something as extreme or more extreme than the observed test statistic given that the null hypothesis is true. Consider a setting in which the null hypothesis is true, for example $\mu_1 = \mu_2$, where $\mu_i$ is the mean value of a trait for individuals with genotype $i$. Now suppose we took 1000 sample sets, where each sample was comprised of $n_1$ individuals from a population with genotype 1 and $n_2$ individuals from a population with genotype 2. If

**Table 4.2.** Errors for multiple hypothesis tests

|       |       | Test |       |           |
|-------|-------|------|-------|-----------|
|       |       | Non-significant | Significant |    |
| Truth | $H_0$ | $U$ | $V$ | $m_0$ |
|       | $H_A$ | $T$ | $S$ | $m - m_0$ |
|       |       | $m - R$ | $R$ | $m$ |

we conducted our test on each of the 1000 sample sets, we would expect the resulting test statistic to be slightly different each time. For example, if we know this statistic has a normal distribution, then the proportion of times the absolute value of the test statistic across the 1000 sample sets is $\geq 1.96$ is 0.05. Thus the $p$-value corresponding to 1.96 is 0.05.

Now consider testing $m$ null hypotheses, given by $H_0^1, \ldots, H_0^m$. Again two types of errors can occur for each of these $m$ tests, as described in Table 4.1. Summing over all of the tests yields the results given by Table 4.2, where $V$ is the number of type-1 errors and $T$ is the number of type-2 errors. The *family-wise error rate (FWER)* is defined as the probability of making at least one type-1 error. Formally, we have

$$FWER = Pr(V \geq 1) \tag{4.1}$$

The FWER is more precisely defined in terms of whether the null hypotheses are indeed true. The FWER under the complete null (FWEC) is the probability that at least one type-1 error occurs given that all null hypotheses are true. That is,

$$FWEC = Pr(V \geq 1 \,|\, H_0^C \text{ true}) \tag{4.2}$$

where $H_0^C = [H_0^1, \ldots, H_0^m]$ is the complete set of all null hypotheses. The FWER under a partial null (FWEP), on the other hand, is conditional on a subset of null hypotheses, say $H_0^{P_1} = [H_0^1, \ldots, H_0^k]$, being true. That is,

$$FWEP = Pr(V \geq 1 \,|\, H_0^{P_1} \text{ true}) \tag{4.3}$$

A test procedure can have strong or weak control of the FWER. A procedure is said to control the FWER at level $\alpha$ in the *weak sense* if the FWEC is less than or equal to $\alpha$. *Strong control* of the FWER is when the FWEP is less than or equal to $\alpha$ under *all* subsets of null hypotheses. That is, strong control implies FWEP$\leq \alpha$ for all partial nulls. Consider for example a simple situation in which we are interested in testing $m = 2$ null hypotheses, given by $H_0^1$ and $H_0^2$. Under the complete null, $m_0$ of Table 4.2 will be given by $m_0 = m = 2$. There are a total of four partial nulls, given by

$$
\begin{aligned}
H_0^{P_1} &= \left[H_0^1, H_0^2\right] \\
H_0^{P_2} &= H_0^1 \\
H_0^{P_3} &= H_0^2 \\
H_0^{P_4} &= \emptyset
\end{aligned}
\tag{4.4}
$$

Control of the FWER in the strong sense means that the FWEP is less than or equal to $\alpha$ for all four of these configurations of true null hypotheses. In general, strong control of the FWER is desirable since we do not know which set of null hypotheses are indeed true.

### 4.1.2 False discovery rate

The *false discovery rate* (FDR) has also gained popularity in the past decade as a measure of error associated with hypothesis testing. Formally, FDR is defined as the expected proportion of null hypotheses that are true among those that are declared significant. Returning to the notation of Table 4.2 and assuming $R$, the number of tests that are declared significant, is greater than 0, FDR is given by

$$
FDR = E\left(\frac{V}{R}\right)
\tag{4.5}
$$

Here $E(\cdot)$ denotes expectation and is defined explicitly in Section 2.1.1. For $R = 0$, we define the ratio $V/R$ as identically equal to 0 since no false rejections can occur. The FDR can therefore be expressed equivalently as

$$
\begin{aligned}
FDR &= E(V/R|R > 0)Pr(R > 0) + E(V/R|R = 0)Pr(R = 0) \\
&= E(V/R|R > 0)Pr(R > 0)
\end{aligned}
\tag{4.6}
$$

In Section 4.2.4, we return to this definition and consider a slight modification, termed the *positive false discovery rate* (pFDR).

There is a well-defined relationship between the FDR and the FWER. To see this, first suppose that all null hypotheses are true. In this case, we have that $V = R$ and

$$
V/R = \begin{cases} 0 \text{ if } V = 0 \\ 1 \text{ if } V \geq 1 \end{cases}
\tag{4.7}
$$

Thus

$$E\,(V/R) = 0 * Pr(V = 0) + 1 * Pr(V \geq 1)$$
$$= Pr(V \geq 1) \tag{4.8}$$
$$= FWER$$

That is, if all null hypotheses are true, then the FDR is equal to the FWER. In other words, we say that control of the FDR leads to control of the FWER in the weak sense. If not all null hypotheses are true, so that $V < R$, then we have $V/R < 1$ and

$$E\,(V/R) = (V/R)\,Pr(V \geq 1) + (0/R)\,Pr(V = 0)$$
$$= (V/R)\,Pr(V \geq 1) \tag{4.9}$$
$$< Pr(V \geq 1)$$

Consequently, we have the general result that the FDR is less than or equal to the FWER. This implies that any approach that controls the FWER will also control the FDR. The reverse, however, is not true. That is, control of the FDR does not generally imply control of the FWER.

Controlling the FDR has become increasingly popular in the context of analyzing a large number of variables. The reason for this stems largely from the fact that many of these hypotheses are expected to be false. As the number of false hypotheses, given by $m - m_0$ in Table 4.2, increases, the number of true positives, given by $S$, will also tend to increase. In turn, $V/R$ will be smaller and the difference between the FDR and the FWER will be greater. The choice of which error measure to use rests heavily on the scientific goal and expectations of our investigation. Consider for example a setting in which the primary aim of the analysis is exploratory in the sense that the discovery of new genes will spark additional confirmatory experiments. In this case, we want very good power to detect associations and making some mistakes is acceptable since we are likely to identify them in subsequent experiments. Controlling the FDR is a natural choice in this setting since it quantifies the proportion of significant tests for which the null is true. On the other hand, if the number of truly false null hypotheses is small or the consequence of incorrectly declaring a test significant is grave, then the FWER may be a more appropriate measure.

## 4.2 Single-step and step-down adjustments

This section includes a discussion of two general types of algorithms for making multiple testing adjustments. The first is the *single-step adjustment*, in which a single criterion is used to assess the significance of all test statistics or corresponding $p$-values. The second is the *step-down adjustment*, which involves ordering test statistics or $p$-values and then using a potentially different criterion for each of the ordered values. The methods of Bonferroni,

Tukey (1977) and Scheffe (1999) are single-step approaches, while the method of Benjamini and Hochberg (1995), described below, is a step-down approach. The resampling procedures of Westfall and Young (1993) and Pollard and van der Laan (2004) can be single-step or step-down and are discussed in Section 4.3.

### 4.2.1 Bonferroni adjustment

Suppose we conduct $m$ tests of hypotheses, given by $H_0^1, \ldots, H_0^m$, and each test is controlled at a level equal to $\alpha$. This means that, for any single test, the probability of incorrectly rejecting the null hypothesis, the type-1 error rate, is less than or equal to $\alpha$. Formally, we write this as

$$Pr(\text{reject } H_0^i \mid H_0^i \text{ true}) \leq \alpha \qquad (4.10)$$

for all $i = 1, \ldots, m$. Now let us assume that these tests are independent of one another, so that the likelihood of rejecting a test does not depend on the outcomes of other tests. If we let $V$ be the number of true nulls that are declared significant, then the probability of incorrectly rejecting at least one null hypothesis is given by

$$
\begin{aligned}
\text{FWEC} &= Pr(V \geq 1 \mid H_0^C \text{ true}) \\
&= 1 - Pr(V = 0 \mid H_0^C \text{ true}) \\
&= 1 - \prod_{i=1}^{m} Pr(\text{do not reject } H_0^i \mid H_0^i \text{ true}) \\
&= 1 - \prod_{i=1}^{m} \left[1 - Pr(\text{reject } H_0^i \mid H_0^i \text{ true})\right] \\
&\leq 1 - \prod_{i=1}^{m} (1 - \alpha) = 1 - (1 - \alpha)^m
\end{aligned}
\qquad (4.11)
$$

Notably, if $m = 1$, then this equation reduces to $FWEC \leq \alpha$. However, if $m = 2$ independent tests are performed, each at level $\alpha = 0.05$, then we are only certain that the probability of making at least one type-1 error is less than or equal to $1 - (1 - 0.05)^2 = 0.0975$. Performing ten tests each at level $\alpha$ only controls the FWEC at a level of $1 - (1 - 0.05)^{10} = 0.401$ and so on. This means that although we are controlling each of the ten individual tests at a level $\alpha$, overall our error may be as great as 40%.

The *Bonferroni* adjustment for multiple comparisons is a single-step procedure and probably the most straightforward adjustment to apply. It involves simply using $\alpha' = \alpha/m$ in place of $\alpha$ for the level of each test, where $m$ is the number of tests to be performed. For example, if we plan to conduct $m = 10$ hypothesis tests and want to control this at an overall level of $\alpha = 0.05$, then we let $\alpha' = 0.05/10 = 0.005$. In this case, Equation (4.11) reduces to

$$\begin{aligned} \text{FWEC} &\leq 1 - (1 - 0.005)^{10} \\ &= 1 - 0.951 = 0.049 \end{aligned} \tag{4.12}$$

That is, if we control each of $m$ tests at the $\alpha/m$ level, then our overall FWEC will be controlled at a level equal to $\alpha$. It is noted that the Bonferroni adjustment is thought to be quite conservative in many genetic investigations since (1) it assumes all tests are independent, which is generally not the case, and (2) it is based on FWER control. As a result, power for detecting associations will be limited. An illustration of applying the Bonferroni adjustment in R is given below.

*Example 4.1 (Bonferroni adjustment).* In this example, we test for associations between mutations in the protease region of the HIV genome and the difference between indinavir (IDV) and nelfinavir (NFV) fold resistance based on the Virco data. Additional information on this dataset is given in Section 1.3.3. Recall that IDV and NFV fold resistance are measured by the variables IDV.Fold and NFV.Fold, respectively, while the categorical variables P1,...,P99 represent the amino acids present at the corresponding protease site. We dichotomize the genotype variables and only consider sites for which at least 5% of individuals have an observed mutation. Also, for simplicity, we assume a missing genotype value is the wildtype genotype.

```
> attach(virco)
> PrMut <- virco[,23:121]!="-" & virco[,23:121]!="."
> NObs <- dim(virco)[1]
> PrMutSub <-data.frame(PrMut[ , apply(PrMut,2,sum) > NObs*.05])
> Trait <- IDV.Fold - NFV.Fold
```

Tests of differences in the trait between individuals with and without a mutation at the corresponding sites are calculated based on the $t$-test. A vector of sorted $p$-values corresponding to these tests is then reported:

```
> TtestP <- function(Geno){
+       return(t.test(Trait[Geno==1],
+       Trait[Geno==0], na.rm=T)$"p.value")
+       }
> Pvec <- apply(PrMutSub, 2, TtestP)
> sort(Pvec)

          P30          P76          P88          P55          P48
3.732500e-12 9.782323e-10 1.432468e-06 2.286695e-06 5.749467e-06
          P89          P11          P82          P60          P85
8.924013e-05 4.171618e-04 9.500604e-04 1.115441e-03 1.219064e-03
```

```
          P54            P43            P61            P46            P67
1.489381e-03 2.025621e-03 2.556156e-03 4.198935e-03 7.765537e-03
          P69            P84            P47            P35            P32
1.113762e-02 1.557464e-02 1.574864e-02 2.392427e-02 2.508445e-02
          P33            P14            P16            P72            P13
2.722251e-02 3.441981e-02 5.570492e-02 5.748494e-02 6.375590e-02
          P15            P34            P53            P64            P90
1.089171e-01 1.167541e-01 1.556130e-01 2.540249e-01 2.618606e-01
          P36            P63            P37            P77            P24
2.896151e-01 2.945370e-01 3.257741e-01 3.356589e-01 3.441678e-01
          P93            P71            P10            P74            P73
3.619516e-01 3.761893e-01 4.268153e-01 4.480744e-01 4.906612e-01
          P20            P19            P58            P62            P41
5.311825e-01 5.342250e-01 5.440101e-01 6.677043e-01 6.998280e-01
          P12            P57
8.050362e-01 9.938846e-01
```

Based on this unadjusted analysis and $\alpha = 0.05$, we would conclude that baseline mutations at each of multiple sites (listed below) are associated with a difference in IDV and NFV fold resistance:

```
> names(PrMutSub)[Pvec < 0.05]
```

```
 [1] "P11" "P14" "P30" "P32" "P33" "P35" "P43" "P46" "P47" "P48" "P54"
[12] "P55" "P60" "P61" "P67" "P69" "P76" "P82" "P84" "P85" "P88" "P89"
```

Bonferroni adjusted $p$-values are generated using the p.adjust() function as follows:

```
> PvecAdj <- p.adjust(Pvec, method="bonferroni")
> sort(PvecAdj)
```

```
          P30            P76            P88            P55            P48
1.754275e-10 4.597692e-08 6.732600e-05 1.074747e-04 2.702250e-04
          P89            P11            P82            P60            P85
4.194286e-03 1.960660e-02 4.465284e-02 5.242573e-02 5.729603e-02
          P54            P43            P61            P46            P67
7.000090e-02 9.520419e-02 1.201393e-01 1.973500e-01 3.649803e-01
          P69            P84            P47            P10            P12
5.234681e-01 7.320083e-01 7.401862e-01 1.000000e+00 1.000000e+00
          P13            P14            P15            P16            P19
1.000000e+00 1.000000e+00 1.000000e+00 1.000000e+00 1.000000e+00
          P20            P24            P32            P33            P34
1.000000e+00 1.000000e+00 1.000000e+00 1.000000e+00 1.000000e+00
          P35            P36            P37            P41            P53
1.000000e+00 1.000000e+00 1.000000e+00 1.000000e+00 1.000000e+00
          P57            P58            P62            P63            P64
1.000000e+00 1.000000e+00 1.000000e+00 1.000000e+00 1.000000e+00
          P71            P72            P73            P74            P77
1.000000e+00 1.000000e+00 1.000000e+00 1.000000e+00 1.000000e+00
```

```
       P90              P93
1.000000e+00  1.000000e+00
```

It is easy to see that this is equivalent to taking the original $p$-values and multiplying by the number of tests, 47 in this case. For example, multiplying the smallest $p$-value, $3.73 \times 10^{-12}$, by 47 yields the adjusted $p$-value of $1.75 \times 10^{-10}$. Adjusted values that are greater than 1 are set equal to 1 since a $p$-value is restricted to the closed set $[0, 1]$. Based on this adjustment, we are only able to reject a subset of the null hypotheses that we rejected previously:

```
> names(PrMutSub)[PvecAdj < 0.05]
```

```
[1] "P11" "P30" "P48" "P55" "P76" "P82" "P88" "P89"
```

### 4.2.2 Tukey and Scheffe tests

Tukey's studentized range test is another single-step adjustment method for testing multiple hypotheses that is useful for the comparison of means between groups (Tukey, 1977). For example, suppose we fit an ANOVA model of the form

$$Y_{ij} = \mu + \alpha_i + \epsilon_{ij} \tag{4.13}$$

where $Y_{ij}$ is the measured trait for the $j$th individual with treatment $i$, $\mu$ is the overall population mean, $\alpha_i$ is the shift in the mean resulting from receiving treatment $i$, $i = 1, \ldots, m$ and $j = 1, \ldots, n_i$. Typically, we begin by conducting an overall F-test for equality of means across the treatment groups. Formally, we test the null hypothesis given by $H_0 : \alpha_1 = \alpha_2 = \ldots = \alpha_m$. If this test is significant, indicating a departure from the null, then interest may lie in identifying the specific treatment groups that are different. That is, we may now want to test the set of "$M$ choose 2" hypotheses, given by $H_0 : \alpha_i = \alpha_j$ for $i \neq j$.

Tukey's honestly significantly different (HSD) test based on the studentized range distribution is a natural approach for this setting. Notably, in its original formulation, this method is applicable to groups that have the same sample size, so we begin by assuming $n_i = n$ for all $i$. For each null hypothesis $H_0 : \mu_i = \mu_j$, a test statistic is given by

$$t_s = \frac{\sqrt{2}\widehat{D}}{\sqrt{\mathrm{Var}[\widehat{D}]}} \sim q_{m,(m \times n) - m} \tag{4.14}$$

where $\widehat{D}$ is the difference in the sample means for groups $i$ and $j$. It can be shown that $t_s$ can also be written as:

$$t_s = \frac{\sqrt{n}\widehat{D}}{\sqrt{MSE}} \sim q_{m,(m \times n)-m} \tag{4.15}$$

where MSE is the mean square error from fitting the full ANOVA model. This statistic has a studentized range distribution with $m$ and $(m \times n) - m$ degrees of freedom. This statistic is also very similar to the statistic we construct in performing a two-sample t-test for the comparison of two independent means. Specifically, $t_s = \sqrt{2}t$, where $t$ is a two-sample $t$-statistic based on equal sample sizes. The adjustment for the number of tests performed comes into play in the degrees of freedom for the studentized range statistic $t_s$. As the numerator degrees of freedom, given by $m$, increase, the critical value $q_{m,(m \times n)-m}$ also increases, and it becomes harder to reject the null hypothesis. In other words, as the number of tests increases, the criterion for rejection becomes more stringent in order to account for chance variation in the test statistics.

In the context of population genetic association studies, we are generally interested in testing whether the trait is the same across all levels of the genotype. For example, suppose our trait is the age of onset of breast cancer and we have a single biallelic candidate SNP taking on the values $A_1A_1$, $A_1A_2$ and $A_2A_2$. If we let $\mu_1$, $\mu_2$ and $\mu_3$ be respectively the population-level mean age of onset of breast cancer for each of these genotypes, the null hypotheses are given by $H_0 : \mu_1 = \mu_2$, $H_0 : \mu_1 = \mu_3$ and $H_0 : \mu_2 = \mu_3$. In this setting, it would be unusual for the sample sizes to be equal across genotypes, and thus an extension of Tukey's method is required. The Tukey-Kramer method involves simply replacing $n$ of Equation (4.15) with the harmonic mean of the two sample sizes, given by $n_{ij} = 2(1/n_i + 1/n_j)^{-1}$. An example of applying this approach in R is provided in the following example.

*Example 4.2 (Tukey's single-step method).* Returning to Example 2.5, we again consider the association between the SNP labeled `resistin_c180g` and the percentage change in the non-dominant muscle strength before and after exercise training, as measured by `NDRM.CH`, using the FAMuSS data. Both the `ptukey()` and `qtukey()` functions in R assume equal sample sizes per group so we employ the `TukeyHSD()` function, which allows for unbalanced data. Recall that without an adjustment for multiple testing, we had the following result:

```
> attach(fms)
> Trait <- NDRM.CH
> summary(lm(Trait~resistin_c180g))

Call:
lm(formula = Trait ~ resistin_c180g)

Residuals:
    Min     1Q  Median     3Q    Max
-56.054 -22.754  -6.054  15.346 193.946
```

```
Coefficients:
                 Estimate Std. Error t value Pr(>|t|)
(Intercept)        56.054      2.004  27.973   <2e-16 ***
resistin_c180gCG   -5.918      2.864  -2.067   0.0392 *
resistin_c180gGG   -4.553      4.356  -1.045   0.2964
---
Signif. codes:  0 *** 0.001 ** 0.01 * 0.05 . 0.1   1

Residual standard error: 33.05 on 603 degrees of freedom
  (791 observations deleted due to missingness)
Multiple R-squared: 0.007296,       Adjusted R-squared: 0.004003
F-statistic: 2.216 on 2 and 603 DF,  p-value: 0.1100
```

We see from this output that the unadjusted Wald test comparing the mean percentage change in muscle strength of individuals with the CG genotype to those with the homozygous wildtype CC genotype yields a significant $p$-value of 0.039. The estimated coefficient of $-5.92$ implies that the mean change in muscle strength for individuals with the CG genotype is lower than the mean change among individuals with the CC genotype. Applying the Tukey approach provides us with adjusted $p$-values:

```
> TukeyHSD(aov(Trait~resistin_c180g))

  Tukey multiple comparisons of means
    95% family-wise confidence level

Fit: aov(formula = Trait ~ resistin_c180g)

$resistin_c180g
            diff        lwr       upr      p adj
CG-CC -5.917630 -12.645660  0.8103998 0.0977410
GG-CC -4.553042 -14.788156  5.6820721 0.5486531
GG-CG  1.364588  -8.916062 11.6452381 0.9478070
```

From this we again see that the difference in means between the individuals with the CG and CC genotypes is $-5.92$; however, we are unable to detect a significant difference in means between genotype pairs after applying the Tukey adjustment for multiple comparisons. The corresponding adjusted $p$-value is 0.098.                                                    □

Another commonly used approach to controlling the FWE is Scheffe's method. This approach differs from Tukey's method in the set of hypotheses considered. Recall that Tukey's method provides an adjustment for testing for differences between all pairs of means. Scheffe's method involves testing a larger set of hypotheses that includes all contrasts of the factor-level means. A *contrast* in the one-way ANOVA setting is defined as a linear combination of the means such that the coefficients sum to zero. More formally, a contrast is written as the function

$$l = \sum_{i=1}^{m} \lambda_i \mu_i \qquad (4.16)$$

such that $\sum_{i=1}^{m} \lambda_i = 0$. For example, in the discussion above, we considered the hypothesis given by $H_0 : \mu_i = \mu_j$, which is equivalent to $H_0 : \mu_i - \mu_j = 0$. This can also be rewritten as $H_0 : \lambda' \mu = 0$, where $\mu = (\mu_1, \ldots, \mu_m)^T$ is the vector of factor means and $\lambda'$ is a vector with a 1 in the $i$th position, a $-1$ in the $j$th position and 0's elsewhere. That is, $\lambda_i = 1$ and $\lambda_j = -1$. The vector $\lambda$ is referred to as the coefficient vector, and we see in this example that the sum of the elements of $\lambda$ is equal to zero. Therefore, we say $l = \lambda' \mu$ is a contrast. There are many different hypotheses about contrasts that we can test. For example, another hypothesis involving a contrast is given by

$$H_0 : \frac{\mu_i + \mu_j}{2} - \frac{\mu_k + \mu_l}{2} = 0, \quad i \neq j \neq k \neq l \qquad (4.17)$$

In this case, we define $\lambda_i = \lambda_j = 1/2$ and $\lambda_k = \lambda_l = -1/2$. We let $\mathcal{L}$ be the set of all linear contrasts of the factor means. As just described, this set includes all pairwise differences in means as well as functions of the form given in Equation (4.17), among others. Scheffe's method is an adjustment approach that controls the FWER when we are interested in testing whether each element of $\mathcal{L}$ is equal to zero.

In order to construct an $F$-test statistic for testing a single contrast $H_0 : \lambda' \mu = 0$, we begin by defining the vector $\rho$ as

$$\rho' = \left( \frac{\lambda_1}{n_1} 1'_{n_1}, \ldots, \frac{\lambda_m}{n_m} 1'_{n_m} \right) \qquad (4.18)$$

where $1_{n_i}$ is an $n_i \times 1$ vector of 1's and $n_i$ is the number of individuals in group $i$. The usual $F$-statistic corresponding to a test of our hypothesis is given by

$$F = \frac{\rho' Y}{MSE(\rho' \rho)} \sim F_{1,(m \times n)-1} \qquad (4.19)$$

where again the MSE is arrived at from fitting an ANOVA model and $Y$ is a vector of observed responses (i.e., the trait of interest). Scheffe's method involves constructing a slightly modified $F$-statistic given by

$$F_s = \frac{\rho' Y / (m-1)}{MSE(\rho' \rho)} \sim F_{m-1,(m \times n)-m} \qquad (4.20)$$

Again the adjustment for multiple comparisons enters into the test statistic through the degrees of freedom. In this case, the numerator degrees of freedom are set equal to $m - 1$, where in the usual setting for testing a single contrast, we set this equal to 1.

Consider for example a situation in which there are $m = 5$ groups, each of size $n = 20$. Suppose an investigator decides to test the null hypothesis $H_0 : \mu_1 = \mu_2$ at the $\alpha = 0.05$ level without making a multiple comparison adjustment. Using Equation (4.19), this investigator calculates the $F$-statistic to be 4.5. Comparing it with the critical value given by $F_{1,40-1} = 4.08$, it is concluded that the null hypothesis is false and indeed there is a difference in the two population means. Note that the denominator degrees of freedom are $40 - 1$ since we are comparing two groups, each of size 20, for this test. Using Scheffe's method, we would instead get $F_s = 4.5/(5-1) = 1.125$ and compare it with the critical value of $F_{4,100-5} = 2.47$. In this case, we would conclude that, based on the observed data, we cannot reject the null hypothesis that the two population means are equal. While a function for implementing Scheffe's test in R is not readily available, it is straightforward to apply this approach using existing functions and the identity derived above. This is given as an exercise.

### 4.2.3 False discovery rate control

As described in Section 4.1.2, the FDR is another intuitively appealing measure of our error rate in the context of genetic investigations. Here we begin by describing one approach to controlling FDR, termed the Benjamini and Hochberg (B-H) adjustment, which is named after the authors who originally proposed it (Benjamini and Hochberg, 1995). Again we begin by considering testing the series of independent null hypotheses given by $H_0^1, \ldots, H_0^m$ and supposing the resulting $p$-values are given by $p_1, \ldots, p_m$. Further, let us suppose that we want to control the false discovery rate at a level $q$. The B-H procedure is summarized in the following three simple steps:

---

ALGORITHM 4.1: FALSE DISCOVERY RATE CONTROL (B-H PROCEDURE):

1. Let $p_{(1)}, \ldots, p_{(m)}$ denote the ordered observed $p$-values such that

$$p_{(1)} \leq \cdots \leq p_{(m)}$$

and let the corresponding null hypotheses be given by $H_0^{(1)}, \ldots, H_0^{(m)}$.

2. Define

$$k = \max \left\{ i : p_{(i)} \leq \frac{i}{m} q \right\} \qquad (4.21)$$

3. Reject $H_0^{(1)}, H_0^{(2)}, \ldots, H_0^{(k)}$.

---

Let us suppose, for example, that we want to test for an association between each of ten SNPs and the presence of disease. For simplicity, we assume that each SNP is in a separate gene and our tests are independent. Further suppose we are primarily interested in the main effects of the SNPs and not their interactions. In this case, for each SNP $i = 1, \ldots, 10$, we construct a $2 \times 3$ contingency table and calculate a $\chi^2$-statistic corresponding to the null hypothesis $H_0 : OR_i = 1$ as described in Section 2.2.1. Suppose the resulting ordered $p$-values are given by

$$0.001 \quad 0.012 \quad 0.014 \quad 0.122 \quad 0.245$$
$$0.320 \quad 0.550 \quad 0.776 \quad 0.840 \quad 0.995$$

The Bonferroni adjustment would lead us to use the adjusted significance level of $\alpha^* = 0.05/10 = 0.005$. Based on this, we would reject only $H_0^{(1)}$. Using the B-H method, we would instead compare the $i$th ordered $p$-value to $\alpha_i^* = 0.05 \, (i/10)$, given by

$$0.005 \quad 0.010 \quad 0.015 \quad 0.020 \quad 0.025$$
$$0.030 \quad 0.035 \quad 0.040 \quad 0.045 \quad 0.050$$

The value of $k$ in Equation (4.21) is simply the maximum $i$ such that $p_{(i)}$ is less than or equal to $\alpha_i^*$. In this example, we have $k = 3$ since $p_{(3)} = 0.014 < 0.015$, while $p_{(i)} > \alpha_i^*$ for $i > 3$. Thus, using the B-H approach, we would reject the null hypotheses $H_0^{(1)}$, $H_0^{(2)}$ and $H_0^{(3)}$. Importantly, the fact that $p_{(2)} = 0.012$ is not less than $\alpha_i^* = 0.010$ is not relevant since the larger $p$-value $p_{(3)} = 0.014$ does meet its rejection criterion. We call this procedure a *step-down adjustment* since each test statistic has a different criterion for rejection.

In addition to defining rejection criteria, we can also calculate adjusted $p$-values. This proceeds in two steps. First we calculate an adjusted $p$ given by $p_{(i)}^{adj} = p_{(i)} m/i$ for each $i$. We then update these $p$-values to ensure monotonicity by letting $p_{(i)}^{adj} = \min_{j \geq i} \left( p_{(j)}^{adj} \right)$. Application of the B-H adjustment and calculation of adjusted $p$-values is illustrated in the following example.

*Example 4.3 (Benjamini and Hochberg (B-H) adjustment).* We return now to the Virco analysis results of Example 4.1. Recall that in that example we calculated a vector of unadjusted $p$-values based on performing multiple $t$-tests and called this Pvec. The B-H adjusted $p$-values are calculated based on this vector using the two-step approach above. In the following code, we first sort from largest to smallest and multiply by the sequence $(m/m, m/(m-1), \ldots, m/2, m/1)$. This is necessary to apply the cummin() function appropriately for monotonicity:

```
> Pvec <- as.vector(Pvec)
> m <- length(Pvec)
> BHp <- sort(Pvec,decreasing=T)*m/seq(m,1)
> sort(cummin(BHp))
```

```
[1]  1.754275e-10 2.298846e-08 2.244200e-05 2.686866e-05 5.404499e-05
[6]  6.990477e-04 2.800943e-03 5.581605e-03 5.729603e-03 5.729603e-03
   ...
[41] 5.946157e-01 5.946157e-01 5.946157e-01 7.132296e-01 7.309314e-01
[46] 8.225370e-01 9.938846e-01
```

The resulting sites that are declared significant based on the B-H adjustment are given as follows. Note that we first reorder the adjusted $p$-values to be consistent with the original ordering:

```
> BHp[order(Pvec,decreasing=T)] <- cummin(BHp)
> names(PrMutSub)[BHp < 0.05]

[1]  "P11" "P30" "P43" "P46" "P47" "P48" "P54" "P55" "P60" "P61" "P67"
[12] "P69" "P76" "P82" "P84" "P85" "P88" "P89"
```

Notably, this is a subset of the sites found based on an unadjusted analysis and is less conservative than the Bonferroni adjustment described in Example 4.1. Finally, the same adjusted $p$-values can also be calculated by specifying method="BH" within the p.adjust() function as follows:

```
> sort(p.adjust(Pvec, method="BH"))

[1]  1.754275e-10 2.298846e-08 2.244200e-05 2.686866e-05 5.404499e-05
[6]  6.990477e-04 2.800943e-03 5.581605e-03 5.729603e-03 5.729603e-03
   ...
[41] 5.946157e-01 5.946157e-01 5.946157e-01 7.132296e-01 7.309314e-01
[46] 8.225370e-01 9.938846e-01                                        □
```

The B-H procedure for controlling the FDR assumes independence of the test statistics corresponding to the true null hypotheses. In a follow-up manuscript, Benjamini and Yekutieli (2001) proved that this procedure will also control the FDR if the test statistics corresponding to the true null hypotheses are positively regression dependent (PRD). The concept of PRD is beyond the scope of this text, though the ambitious reader is encouraged to read the cited manuscript for specific examples of when PRD holds. Benjamini and Yekutieli (2001) also propose an extension of the B-H approach that controls the FDR in settings for which PRD does not hold. This extension is to simply replace $q$ of Equation (4.21) with the quantity $\tilde{q} = q/\sum_{i=1}^{m}(1/i)$ and is referred to as the Benjamini and Yekutieli (B-Y) adjustment. Application of this adjustment is illustrated in the following example.

*Example 4.4 (Benjamini and Yekutieli adjustment).* Again using the vector of $p$-values from Example 4.1, the B-Y adjustment is calculated using the p.adjust() function as follows:

```
> BYp <- p.adjust(Pvec, method="BY")
> sort(BYp)
```

```
[1]  7.785410e-10 1.020220e-07 9.959678e-05 1.192422e-04 2.398497e-04
[6]  3.102348e-03 1.243048e-02 2.477096e-02 2.542777e-02 2.542777e-02
     ...
[41] 1.000000e+00 1.000000e+00 1.000000e+00 1.000000e+00 1.000000e+00
[46] 1.000000e+00 1.000000e+00
```

The resulting $p$-values are more conservative than we saw with the application of the B-H approach in Example 4.3. The sites corresponding to significant associations are now given by

```
> names(PrMutSub)[BYp < 0.05]
```

```
[1]  "P11" "P30" "P43" "P48" "P54" "P55" "P60" "P61" "P76" "P82" "P85"
[12] "P88" "P89"
```
□

### 4.2.4 The $q$-value

The *q-value* is an alternative measure of significance based on the FDR concept that was recently proposed for genome-wide association studies (Storey, 2002, 2003; Storey and Tibshirani, 2003). Before defining the $q$-value, we first introduce the *positive false discovery rate* (pFDR). The pFDR is defined formally as

$$pFDR = E\left[\frac{V}{R}\middle|R > 0\right] \qquad (4.22)$$

and differs from the FDR by the multiplicative factor $Pr(R > 0)$, as seen by returning to Equation (4.6). A primary motivation for this alternative quantity is the intuitive appeal of conditioning on the occurrence of at least one positive finding. Notably, however, the pFDR cannot be controlled in a traditional sense since it is identically equal to 1 if the proportion of true null hypotheses, given by $m_0/m$ in Table 4.2, is equal to 1. That is, we cannot guarantee that pFDR$\leq \alpha$ for $\alpha < 1$. The FDR, on the other hand, is not subject to this limitation. The $q$-value is based on an alternative paradigm that eliminates this concern. Specifically, rather than fixing an error rate and then estimating a significance threshold that maintains this rate on average, we fix the significance threshold and then estimate the rate over that threshold. The familiar *significance analysis of microarrays* (SAM) similarly involves fixing rejection regions and then estimating the corresponding FDRs (Tusher *et al.*, 2001).

Recall that we have defined the $p$-value as the probability of seeing something as extreme as or more extreme than the observed test statistic given that the null hypothesis is true. Formally, we can write the $p$-value, corresponding to an observed test statistic $T = t$ as

$$p(t) = \inf_{\{\Gamma : t \in \Gamma\}} Pr\left(T \in \Gamma | H_0\right) \qquad (4.23)$$

where inf ("infimum") is defined as the greatest lower bound over the corresponding set and $\{\Gamma : t \in \Gamma\}$ is a set of nested rejection regions that contain the observed test statistic $t$. For example, we can define $\{\Gamma\} = [c, \infty)$ for all real-valued $c < t$. That is, we can think of the $p$-value as the minimum probability under the null that our test statistic is in the rejection region (i.e., the minimum type-1 error rate) over the set of nested rejection regions containing the observed test statistic. A set of nested rejection regions for $p$-values, on the other hand, might be given by $\{\Gamma\} = \{[0, \gamma] : \gamma \geq 0\}$. The reader is referred to Lehmann (1997) for more details.

The $q$-value is defined similarly as the greatest lower bound of the pFDR that can result from rejecting a test statistic based on a rejection region in the set of nested rejection regions. Formally, the $q$-value is written

$$q(t) = \inf_{\{\Gamma : t \in \Gamma\}} [\text{pFDR}(\Gamma)] \tag{4.24}$$

Similar to a $p$-value, the $q$-value can be thought of as the expected proportion of false positives among all features that are as extreme as or more extreme than the feature under consideration. In the context of SAM, the $q$-value is computed as the FDR for the smallest estimated rejection region for which the gene under consideration is called significant (Chu *et al.*, 2008). Intuitively, the $q$-value is simply the minimum pFDR that can occur over the set of nested rejection regions when we reject our test statistic $T = t$. This parallels nicely with the definition of $p$-value as the minimum type-1 error rate that can occur over all nested rejection regions when we reject our test statistic. For the interested reader, a well-formulated Bayesian interpretation of the $q$-value as a posterior $p$-value can be found in Storey (2003).

Similar to FDR control, use of the $q$-value is most appropriate when the number of tests performed, given by $m$ in Table 4.2, is large. In this case, the probability that at least one test is declared significant, given by $Pr(R > 0)$, is close to 1. If we set the tuning parameter $\lambda$, a quantity that informs us about the proportion of true null hypotheses, equal to 0, then the $q$-value results in the same adjusted $p$-values as the FDR adjustment described in Section 4.2.3; however, this is a conservative estimate of the $q$-value, and optimizing the choice of $\lambda$ is tenable. Calculation of the $q$-value in R is illustrated in the following example using the `qvalue()` function within the `qvalue` package.

*Example 4.5 (Calculation of the q-value).* The $q$-value is calculated using the `qvalue()` function in the `qvalue` package. Again using the vector of $p$-values from Example 4.1 and setting the tuning parameter $\lambda$ equal to 0, we get the same adjusted $p$-values as we saw in Example 4.3:

```
> library(qvalue)
> sort(qvalue(Pvec,lambda=0)$qvalues)

[1] 1.754275e-10 2.298846e-08 2.244200e-05 2.686866e-05 5.404499e-05
[6] 6.990477e-04 2.800943e-03 5.581605e-03 5.729603e-03 5.729603e-03
```

```
    . . .
[41] 5.946157e-01 5.946157e-01 5.946157e-01 7.132296e-01 7.309314e-01
[46] 8.225370e-01 9.938846e-01
```

If instead of specifying $\lambda$ we specify to use the bootstrap estimation method, pi0.method="bootstrap", we get less conservative estimates of the $q$-values, as shown below:

```
> sort(qvalue(Pvec,pi0.method="bootstrap")$qvalues)

 [1] 2.488334e-11 3.260774e-09 3.183262e-06 3.811158e-06 7.665956e-06
 [6] 9.915570e-05 3.972969e-04 7.917170e-04 8.127096e-04 8.127096e-04
    . . .
[41] 8.434265e-02 8.434265e-02 8.434265e-02 1.011673e-01 1.036782e-01
[46] 1.166719e-01 1.409765e-01
```

In this case, many more sites appear to be significant predictors of a difference in IDV and NFV fold resistance. The qvalue() function can also give us an estimate of the proportion of true null hypotheses, given by $m_0/m$, where $m_0$ and $m$ are respectively the number of true null hypotheses and total hypotheses, as defined in Table 4.2. In this example, this proportion is given by

```
> qvalue(Pvec,pi0.method="bootstrap")$pi0
```

```
[1] 0.1418440
```                                                                    □

## 4.3 Resampling-based methods

Resampling-based methods are an alternative to the single-step and step-down procedures described above that involves taking repeated samples from the observed data. One primary advantage of resampling-based methods is that they offer a natural approach to account for underlying unknown correlation structure among multiple hypotheses. We begin by describing one popular approach given by Westfall and Young (1993), termed the *free step-down resampling (FSDR) method*, and how it can be applied in the context of a population-based investigation with covariates. This approach makes one strong assumption, called *subset pivotality*, which may or may not be appropriate in the settings under consideration. We thus also discuss an alternative approach proposed by Pollard and van der Laan (2004) that relaxes this assumption.

### 4.3.1 Free step-down resampling

Suppose again that we are interested in testing a series of $m$ null hypotheses denoted $H_0^1, \ldots, H_0^m$. For example, the $j$th null hypothesis may be that there is no association between the $j$th SNP under investigation and a marker for disease progression. The idea behind the resampling-based approach we describe is that by taking repeated samples of the observed data, we can simulate the distribution of the test statistics (or $p$-values) under the complete null hypothesis, $H_0^C$. Recall that the complete null is defined by

$$H_0^C = H_0^1 \cap H_0^2 \cap \ldots \cap H_0^m \tag{4.25}$$

where $\cap$ denotes intersection. That is, the complete null refers to the situation in which all null hypotheses are true. We then compare the observed test statistics to this empirical distribution to ascertain the corresponding significance of our tests. The *subset pivotality* condition states that the distribution of test statistics is the same under any combination of true null hypotheses. That is, the test statistic distribution is invariant to whether all null hypotheses are indeed true $(H_0^C)$ or a partial set of null hypotheses are true. Specifically, the covariance between test statistics is assumed to be the same under all scenarios of true and false null hypotheses. Under this assumption, importantly, error control under the complete null will give us the desired control under the true data-generating distribution. The FSDR approach hinges on this assumption and is described first for a quantitative trait. We then discuss application of this algorithm to a binary trait and methods for incorporating confounders and effect modifiers into our analysis.

The FSDR method proceeds in three steps, as follows:

ALGORITHM 4.2: FREE STEP-DOWN RESAMPLING (MAXT):

1. Determine the "observed" test statistics and $p$-values. We begin by letting $x_j$ represent our genotype variables for $j = 1, \ldots, m$ and suppose the phenotype under study is given by $y$. Based on the observed data, we can construct the linear model

$$y_i = \beta_0 + \beta_1 x_{i1} + \ldots + \beta_m x_{im} + \epsilon_i \tag{4.26}$$

for $i = 1, \ldots, n$, where $n$ is our sample size and we assume $\epsilon_i \sim N(0, \sigma^2)$. Alternative formulations of this initial model are discussed below. Using ordinary least squares regression, we arrive at an estimate, $\widehat{\beta}$, of the vector of parameters $\beta = (\beta_0, \beta_1, \ldots, \beta_m)^T$, as described in Chapter 2. In addition, for each $\beta_j$, we can construct a test statistic and $p$-value, given by $T_j$ and $p_j$, respectively, corresponding to the null hypothesis of $H_0 : \beta_j = 0$. For example, $T_j$ may be a Wald test statistic as described in Section 2.2.3, and $p_j = Pr(|T_j| > t_{(n-1),(1-\alpha)/2})$. We term these the "observed" test statistics and $p$-values since they are based on the original data. Now let the corresponding ordered absolute value of the test statistics, sorted from smallest to largest, be given by $|T|_{(1)}, \ldots, |T|_{(m)}$.

2. Generate the (approximate) distribution of test statistics under the complete null. The next step involves determining the distribution of the ordered test statistics under the complete null hypothesis. To do this for

our setting, we begin by determining the residuals from the model-fitting procedure described in STEP 1. These are given by $\widehat{r}_i = y_i - \begin{bmatrix} 1 & \mathbf{x}_i^T \end{bmatrix} \widehat{\beta}$, where $\mathbf{x}_i^T = (x_{i1}, \ldots, x_{im})$. Now we sample with replacement from these residuals to get a bootstrap dataset. That is, for $i = 1, \ldots, n$, we let $y_i^* = \widehat{r}_i^*$, where $\widehat{r}_i^*$ is drawn with replacement from the original set of residuals $\widehat{r}_1, \ldots, \widehat{r}_n$. Using these new data, $y_1^*, \ldots, y_n^*$ as our response and the original design matrix $X$, we refit the model given in Equation (4.26) and determine corresponding test statistics. The absolute values of these are recorded as $|T|_{(1)}^*, \ldots, |T|_{(m)}^*$, where the ordering is the same as the ordering of the original test statistics. Note that at this step the $|T|^*$ are not necessarily ranked from smallest to largest, or, in other words, *monotonicity* does not hold. These resulting test statistics are one realization from the complete null generating distribution.

3. Compare the observed test statistics to test statistics under the complete null to get adjusted $p$-values. The resampling component of STEP 2 above is repeated $B$ times to arrive at multiple bootstrap samples. For each sample, successive maxima are defined as

$$q_1^* = |T|_{(1)}^*$$
$$q_2^* = \max(q_1^*, |T|_{(2)}^*)$$
$$q_3^* = \max(q_2^*, |T|_{(3)}^*) \tag{4.27}$$

$$\vdots$$

$$q_m^* = \max(q_{(m-1)}^*, |T|_{(m)}^*)$$

and we determine whether $q_j^* > |T|_{(j)}$. That is, we check whether the $j$th ordered test statistic is less than the corresponding statistic that was generated based on the distribution of test statistics under the complete null. The adjusted $p$-values, given by $\tilde{p}_{(j)}$ for $j = 1, \ldots, m$, are then defined as the proportion of the $B$ bootstrap samples for which this inequality holds. More formally, we write

$$\tilde{p}_{(j)} = \frac{1}{B} \sum_{b=1}^{B} I\left(q_j^{*(b)} > |T|_{(j)}\right) \tag{4.28}$$

where $I(\cdot)$ is the indicator function, which equals 1 if the argument is true and 0 otherwise, and $b$ indicates the specific bootstrap sample. Finally, monotonicity of these resulting adjusted $p$-values is ensured by completing this final step:

$$\tilde{p}^*_{(m)} = \tilde{p}^*_{(m)}$$
$$\tilde{p}^*_{(m-1)} = \max(\tilde{p}^*_{(m)}, \tilde{p}^*_{(m-1)})$$
$$\vdots$$
$$\tilde{p}^*_{(1)} = \max(\tilde{p}^*_{(2)}, \tilde{p}^*_{(1)})$$

$$(4.29)$$

Notably, Westfall and Young (1993) demonstrate that under the subset pivotality assumption, this approach controls the FWE in the strong sense. This approach is also referred to as the $maxT$ procedure. Replacing the test statistics of Equation (4.27) in step (3) with $p$-values and taking successive minima yields the $minP$ approach to multiple testing. In the following example, we demonstrate this procedure through direct coding. The mt.maxT() and mt.minP() functions in the multtest package can also be used to implement the FSDR approach. However, application of these functions, originally written for gene expression data, to data arising from SNP association studies is not straightforward. To see this, recall that, in its simplest form, expression data generally include quantitative measurements for each individual on each of multiple genes and a single class label, such as disease status. In this case, interest lies in multiple tests of association between the presence of the disease and gene expression levels across several genes. In our setting, we instead have a single quantitative or binary trait measured for each individual and multiple class labels, representing each of the SNPs under investigation. The multiple tests of interest thus correspond to tests of association between the trait and each class label.

*Example 4.6 (Free step-down resampling adjustment).* Returning to the FA-MuSS data, we consider whether there is an association between change in muscle strength of the non-dominant arm and the presence of two variant alleles for each of the four SNPs within the actn3 gene. Let us begin by fitting a multivariable linear model that includes four indicators for the presence of two variant alleles at each of the corresponding SNPs:

```
> attach(fms)
> Actn3Bin <- data.frame(actn3_r577x!="TT",actn3_rs540874!="AA",
+               actn3_rs1815739!="TT",actn3_1671064!="GG")
> Mod <- summary(lm(NDRM.CH~.,data=Actn3Bin))
> Mod

Call:
lm(formula = NDRM.CH ~ ., data = Actn3Bin)

Residuals:
    Min     1Q  Median     3Q     Max
-55.181 -22.614  -7.414  15.486 198.786

Coefficients:
```

```
                       Estimate Std. Error t value Pr(>|t|)
(Intercept)              54.700       3.212  17.028  <2e-16 ***
actn3_r577x.....TT.TRUE  -12.891      4.596  -2.805  0.0052 **
actn3_rs540874.....AA.TRUE 10.899    11.804   0.923  0.3562
actn3_rs1815739.....TT.TRUE 27.673   17.876   1.548  0.1222
actn3_1671064.....GG.TRUE -29.166    17.516  -1.665  0.0964 .
---
Signif. codes:  0 *** 0.001 ** 0.01 * 0.05 . 0.1   1

Residual standard error: 32.93 on 591 degrees of freedom
  (801 observations deleted due to missingness)
Multiple R-squared: 0.01945,       Adjusted R-squared: 0.01281
F-statistic:  2.93 on 4 and 591 DF,  p-value: 0.02037
```

The first step of the FSDR approach is to record the "observed" test statistics. We see from the output above that these statistics are given by $-2.81, 0.923, 1.55$ and $-1.67$. We take the absolute values of these statistics and record the ordered values as follows:

```
> TestStatObs <- Mod$coefficients[-1,3]
> Tobs <- as.vector(sort(abs(TestStatObs)))
```

Before applying the resampling procedure, we need to subset the data that went into the analysis above. Recall from the modeling output that $n = 801$ observations were deleted due to missingness in the genotype or trait variables. We also need to record the ordering of our original test statistics. We do these two steps as follows:

```
> MissDat <- apply(is.na(Actn3Bin),1,any) | is.na(NDRM.CH)
> Actn3BinC <- Actn3Bin[!MissDat,]
> Ord <- order(abs(TestStatObs))
```

The second step of the FSDR requires resampling from the residuals and arriving at test statistics under the null generating distribution. This is achieved using the following for loop:

```
> M <- 1000
> NSnps <- 4
> Nobs <- sum(!MissDat)
> TestStatResamp <- matrix(nrow=M, ncol=NSnps)
> for (i in 1:M){
+       Ynew <- sample(Mod$residuals, size=Nobs, replace=T)
+       ModResamp <- summary(lm(Ynew~., data=Actn3BinC))
+       TestStatResamp[i,] <- abs(ModResamp$coefficients[-1,3])[Ord]
+       }
```

We see that in each iteration of the for loop we (1) take a sample from the model residuals with replacement, (2) refit the model using this sample as the new outcome and (3) record the test statistics corresponding to our ordered observed statistics, given by `Tobs`. The result is a matrix of test statistics, called `TestStatResamp`, corresponding to the null distribution. The final step is to compare our observed test statistics with the distribution of test statistics we generated. To do this, we first arrive at successive maxima by applying the `cummax()` function to each row of our matrix of resampling-based statistics:

```
> Qmat <- t(apply(TestStatResamp, 1, cummax))
```

Adjusted $p$-values are then given by

```
# Note that your code will result in slightly different
# values since we took a random sample above.

> Padj <- apply(t(matrix(rep(Tobs,M), NSnps)) < Qmat, 2, mean)
> Padj
```

```
[1] 0.310 0.203 0.203 0.034
```

Here monotonicity of the resulting $p$-values is already achieved and an additional step is not needed. Based on this analysis, we conclude that individuals who have at least one variant allele at the `actn_577x` SNP have a significantly lower percentage change in non-dominant arm muscle strength than individuals who are homozygous wildtype at this SNP (adjusted $p = 0.034$). Notably, this analysis controls for other SNPs within the `actn` gene but not additional potential predictors, such as race/ethnicity, age and gender. Inclusion of these variables in the initial multivariable model is straightforward. Depending on whether interest lies in testing hypotheses relating to these variables as well, the set of recorded test statistics can be limited to the SNPs under study or expanded to include additional covariates in the model.     □

A slight modification of the algorithm above is required in the context of a binary trait, such as an indicator for the presence of disease. In this case, instead of the linear regression model described in Equation (4.26), we can fit a logistic model of the form

$$\text{logit}(\pi_i) = \beta_0 + \beta_1 x_{i1} + \ldots + \beta_m x_{im} \tag{4.30}$$

where $\pi_i = Pr(y_i = 1 | x_i)$, as described in Section 2.2.3. In turn, rather than resampling from the set of model residuals, we generate binary $y_i^*$ such that

$$y_i^* = \begin{cases} 1 \text{ with probability } \widehat{\pi}_i \\ 0 \text{ with probability } 1 - \widehat{\pi}_i \end{cases} \tag{4.31}$$

where $\widehat{\pi}_i = \exp(\mathbf{x}_i^T \widehat{\boldsymbol{\beta}})/(1 + \exp(\mathbf{x}_i^T \widehat{\boldsymbol{\beta}}))$ for $i = 1, \ldots, n$. With this modification to the resampling component of step 2, the same approach as described above can be applied to this setting. Importantly, however, the subset pivotality condition is not met for this setting, as described in Chapter 6 of Westfall and Young (1993). Application of the FSDR algorithm to this binary trait setting is left as an exercise.

### 4.3.2 Null unrestricted bootstrap

As described above, the free step-down approach hinges on the assumption that the test statistic null distribution can be validly determined by resampling data under the complete null. This holds under the subset pivotality condition, which states that the distribution of test statistics for a subset of true null hypotheses is the same regardless of whether just this subset is true or the complete null is true. However, if this assumption is violated, an alternative multiple testing procedure is needed. Here we describe an approach proposed by Pollard and van der Laan (2004), termed the null unrestricted bootstrap approach, for determining the null distribution of the test statistics that does not require the subset pivotality assumption. Recall that for the free step-down approach, we resample data from the complete null distribution and then generate the test statistic distribution based on these resampled data. Now we will instead arrive at the test statistic distribution based on the original data. In turn, the projection of this distribution onto the space of mean zero distributions yields asymptotic strong control of the FWE.

More formally, let $Q_{0n}$ be the complete null distribution of test statistics. This is the distribution that we aim to determine. Further suppose $P_0$ is the distribution of the data under the complete null and $P$ is the true data distribution. Now suppose $Q_n(P_0)$ is the distribution of test statistics under the null generated distribution. For example, $Q_n(P_0)$ may be the distribution of test statistics that we arrive at by first resampling from the residuals resulting from a model-fitting procedure, as described in Section 4.3.1. Now consider the simple example in which this distribution of test statistics based on the complete null generated distribution is given by $Q_n(P_0) = MVN(0, \Sigma(P_0))$. If subset pivotality does not hold and specifically $\Sigma(P_0)$ does not equal $\Sigma(P)$, then $Q_n(P_0)$ will not equal $Q_{0n} = MVN(0, \Sigma(P))$, the distribution we aim to determine. The basic idea behind the null unrestricted bootstrap approach is to project the distribution of test statistics, given by $Q_n(P)$ for the true data-generating distribution, onto the space of mean zero distributions to arrive at $Q_{0n}$.

To see how this approach works in practice, consider again the linear model given in Equation (4.26). Suppose we are interested in testing the series of null hypotheses given by $H_0 : \beta_j = 0$ for $j = 1, \ldots, m$. The null unrestricted bootstrap approach proceeds as follows:

---

ALGORITHM 4.3: NULL UNRESTRICTED BOOTSTRAP:

1. Fit the model of Equation (4.26) and calculate the least squares estimate of $\boldsymbol{\beta}$, denoted $\widehat{\beta}_n$.

2. Bootstrap $(y_i, \mathbf{x}_i)$ with replacement, preserving the within-individual link.

3. Estimate $\boldsymbol{\beta}$ based on the bootstrap sample and denote this $\beta_n^{\#}$, with the $j$th element corresponding to the $j$th hypothesis under study.

4. Record the vector of statistics

$$Z_n^{\#1} = (\beta_n^{\#} - \widehat{\beta}_n)/sd(\beta_n^{\#}) \qquad (4.32)$$

5. Repeat steps (2)–(4) $B$ times to get $Z_n^{\#1}, \ldots, Z_n^{\#B}$. The distribution of $Z_n^{\#}$, given by $Q_{0n}^{\#}$, converges to $Q_{0n}$ conditional on the data.

6. Determine a single-step significance cutoff by choosing a vector $c = (c_1, \ldots, c_m)$ such that

$$Pr\left[\sum_{j=1}^{m} I\left\{|Z_{jn}^{\#}| > c_j\right\} \geq k\right] = \alpha \qquad (4.33)$$

where $Z_{jn}^{\#}$ is the $j$th element of $Z_n^{\#}$. Here $\alpha$ is the level at which we want to control the type-1 error, and typically we let $k = 1$.

---

This approach is demonstrated in the following example.

*Example 4.7 (Null unrestricted bootstrap approach).* Returning to the data setting and model from Example 4.6, we now apply the null unrestricted bootstrap approach. This begins by defining the estimated coefficients based on the model denoted Mod:

```
> CoefObs <- as.vector(Mod$coefficients[-1,1])
```

The following for loop is then applied, where Nobs, MissDat and Actn3BinC are defined in Example 4.6:

```
> B <-1000
> TestStatBoot <- matrix(nrow=B,ncol=NSnps)
> for (i in 1:B){
+       SampID <- sample(1:Nobs,size=Nobs, replace=T)
+       Ynew <- NDRM.CH[!MissDat][SampID]
+       Xnew <- Actn3BinC[SampID,]
+       CoefBoot <- summary(lm(Ynew~.,data=Xnew))$coefficients[-1,1]
+       SEBoot <- summary(lm(Ynew~.,data=Xnew))$coefficients[-1,2]
```

```
+          if (length(CoefBoot)==length(CoefObs)){
+               TestStatBoot[i,] <- (CoefBoot-CoefObs)/SEBoot
+               }
+          }
```

We see here that we begin by drawing a bootstrap sample from the data (both the trait and genotypes) without disrupting the within-individuals link. We then fit a model based on these data and calculate the vector of test statistics, $Z_n^{\#}$. Notably, if there are insufficient data in the bootstrap sample to fit the full model (that is, estimate all four coefficients), we assume this is non-informative and do not record the results. This occurred in 11 of the 1000 bootstrap samples. Finally, we determine a significant threshold as follows:

```
> for (cj in seq(2.7,2.8,.01)){
+       print(cj)
+       print(mean(apply(abs(TestStatBoot)>cj,1,sum)>=1,na.rm=T))
+       }
# Note that, depending on your sample,
# a different range for cj may be required

[1] 2.7
[1] 0.06471183
[1] 2.71
[1] 0.06268959
[1] 2.72
[1] 0.05965622
[1] 2.73
[1] 0.0586451
[1] 2.74
[1] 0.05662285
[1] 2.75
[1] 0.05460061
[1] 2.76
[1] 0.05257836
[1] 2.77
[1] 0.05055612
[1] 2.78
[1] 0.04954499
[1] 2.79
[1] 0.04954499
[1] 2.8
[1] 0.04853387
```

From this output, we see that a significance threshold of 2.78 for all four tests maintains a type-1 error rate of less than 0.05. Comparing this with the observed test statistics given in Example 4.6, we again conclude that the actn3_577x SNP is significantly associated with percentage change in the non-dominant arm muscle strength based on the multivariable model under consideration.                                                                        □

## 4.4 Alternative paradigms

In Sections 4.1– 4.3, we focused on measures of error and approaches to adjusting for multiple comparisons. In this section, we shift our focus to methods for reducing the number of tests. First, we present the approach, first described by Cheverud (2001), that takes into account LD structure to determine the *effective number of tests* ($M_{\text{eff}}$). Second, we describe a global testing framework, proposed independently by Goeman *et al.* (2004) for gene expression data and Foulkes *et al.* (2005) for SNP data, that obviates the need for a multiple testing adjustment in some settings.

### 4.4.1 Effective number of tests

As described in Section 4.2.1, the Bonferroni adjustment is conservative in the context of many genetic association studies since the SNPs under investigation are usually correlated with one another. This serves as the primary motivation behind the *effective number of tests* ($M_{\text{eff}}$) approach described by Cheverud (2001), Nyholt (2004), Li and Ji (2005) and Gao *et al.* (2008). This approach has gained in popularity in recent years due to its intuitively appealing interpretation and ease of implementation.

Briefly, this approach draws on the fact that the variance of the *eigenvalues* of a correlation matrix of a set of variables captures information on the collective correlation of this set. An eigenvalue is a characteristic of a matrix that is defined as the amount by which a vector in the direction of the corresponding eigenvector is stretched or shrunken when acted upon by this matrix. Formally, we write

$$Ax = \lambda x \tag{4.34}$$

where $x$ is an eigenvector of the matrix $A$ and $\lambda$ is the corresponding eigenvalue. If all of the variables are perfectly correlated with one another, so that the correlation matrix is given by

$$V_{M \times M} = \begin{pmatrix} 1\,1 \ldots 1 \\ 1\,1 \ldots 1 \\ \vdots \qquad \vdots \\ 1\,1 \ldots 1 \end{pmatrix} \tag{4.35}$$

we will have a single non-zero eigenvalue, given by $M$, where $M$ is the number of variables in our set. In this case, the variance of the eigenvalues, given by $\text{Var}[M, 0, 0, \ldots, 0]$, is equal to $M$. On the other hand, if the correlation is 0 between all pairs of observations so that

$$V_{M \times M} = \begin{pmatrix} 1\,0\,\ldots\,0 \\ 0\,1\,\ldots\,0 \\ \ddots \\ 0\,0\,\ldots\,1 \end{pmatrix} \tag{4.36}$$

then all of the eigenvalues will be equal to 1 and the variance of the eigenvalues is 0. Thus the variance of the eigenvalues of the correlation matrix of $M$ variables ranges from 0 to $M$ and depends on the amount of pairwise correlation between variables.

The proportion reduction in the number of tests due to correlation is then characterized by the ratio $\mathrm{Var}[\lambda_{obs}]/M$, where $\lambda_{obs}$ is the set of observed eigenvalues for the correlation matrix of the $M$ variables in our set. The effective number of tests ranges from 1 to $M$ and is defined by

$$M_{\mathrm{eff}} = 1 + (M-1)\left[1 - \frac{Var(\lambda_{obs})}{M}\right] \tag{4.37}$$

For example, if all of the variables are perfectly correlated, as described by Equation (4.35), then $\mathrm{Var}[\lambda_{obs}] = M$ and $M_{\mathrm{eff}} = 1$. On the other hand, if we have no correlation among the set of SNPs under investigation, as given by Equation (4.36), then $\mathrm{Var}[\lambda_{obs}] = 0$ and $M_{\mathrm{eff}} = M$. In a recent manuscript, Gao et al. (2008) propose using an alternative definition for the effective number of tests, given by $M_{\mathrm{eff\text{-}G}}$ and defined as the number of principal components of the correlation matrix that explain $C\%$ of the variability in the data. More formally, $M_{\mathrm{eff\text{-}G}}$ is the minimum $x$ such that $\sum_{i=1}^{x} \lambda_i / \sum_{i=1}^{M} \lambda_i > C$, where $C$ is a predefined threshold. Additional details on principal components analysis (PCA) can be found in Section 3.3.3.

Once the effective number of tests is determined, the resulting value can be used in place of $M$ in our usual adjustment of multiple testing. For example, application of the Bonferroni adjustment is straightforward, where we now use

$$\alpha'' = 1 - (1 - \alpha)^{1/M_{\mathrm{eff}}} \approx \alpha/M_{\mathrm{eff}} \tag{4.38}$$

as our significance threshold. The results of this analysis may be sensitive to the choice of correlation matrix. In the original formulation of this approach, Cheverud (2001) proposes defining genotype scores ($+1$, 0 and $-1$) based on the observed genotype at each marker and then calculating Pearson's product-moment correlation, defined in Section 2.2.1. Gao et al. (2008) propose applying a composite LD between markers (see, for example, Weir (1996)), which reduces to the same Pearson correlation coefficient using a 0, 1, 2 coding of genotypes. Alternatively, Nyholt (2004) proposes using the measure of LD given by $r^2$ of Section 3.1.1 as elements of the correlation matrix. In all cases, a spectral decomposition of the correlation matrix yields the corresponding eigenvalues. Notably, for population-based association studies, allelic phase is potentially unobservable and derivation of $r^2$ thus requires consideration of this missing data challenge, as described in Section 3.1.1.

We leave application of the $M_{\text{eff}}$ approach as an exercise for the reader. This section is included in this text due to the increasing popularity of the approach. Notably, however, this approach has met some criticism in recent reports. Specifically, Salyakina *et al.* (2005) suggest that using either $r^2$ or Pearson's correlation as a measure of correlation in the application of this method is anti-conservative in the context of high LD, as demonstrated through simulation studies. Further characterization and development of this approach is warranted and will likely serve to elucidate its usefulness.

### 4.4.2 Global tests

In Sections 2.2.2 and 4.2.2, we briefly describe the analysis of variance (ANOVA) as an analytic approach for characterizing and testing the equality of means across multiple groups. This approach can be thought of as a global testing framework since it allows us to determine simultaneously if there is a difference across genotypes. For example, suppose we have three observed genotypes, given by $AA$, $Aa$ and $aa$, and we aim to determine whether the mean cholesterol level is the same across these genotypes. We can construct for this setting the model

$$y_{ij} = \mu + \alpha_i + \epsilon_{ij} \tag{4.39}$$

where $y_{ij}$ is the observed trait (cholesterol level in our example) for the $j$th individual with genotype $i$ and $i = 1, 2, 3$. Here $\mu$ is the overall population mean, $\alpha_i$ is the deviation from this mean for individuals with genotype $i$ and $\epsilon_{ij} \sim N(0, \sigma^2)$ is the associated error in the observed data.

An overall test for association between genotype and traits can be expressed as a test of the null hypothesis:

$$H_0 : \alpha_i = 0 \text{ for } i = 1, 2, 3 \tag{4.40}$$

A corresponding test statistic is given by the ratio of the within- and between-group variabilities and has an $F$-distribution with $p - 1$ and $n - p$ degrees of freedom, where $p$ is the number of genotype levels and $n$ is our sample size. Specifically, we have

$$F = \frac{\text{BSS}/(p - 1)}{\text{WSS}/(n - p)} \sim F_{p-1, n-p} \tag{4.41}$$

under $H_0$, where BSS is the between group sum of squares and WSS is the within group sum of squares. Additional details on this statistic, including its derivation, can be found in Rosner (2006) (algebraic representation) and Christensen (2002) (matrix representation).

Now suppose we are interested in testing a group of genes or multiple SNPs within a single gene. We can construct several models and corresponding hypotheses that may be relevant to our investigation. For example, suppose we have two SNPs within a single gene so that there are $3^2 = 9$ possible genotypes, given by

$$G = \{(AA, BB), (AA, Bb), (AA, bb),$$
$$(Aa, BB), (Aa, Bb), (Aa, bb), \qquad (4.42)$$
$$(aa, BB), (aa, Bb), (aa, bb)\}$$

Again we can consider the model given by Equation (4.39), where now we have nine genotypes, or genotype groups, so that $i = 1, 2, \ldots, 9$. In this case, the test statistic corresponding to a test that all of the genotype group effects are 0 (that is, a test of $H_0 : \alpha_i = 0$ for $i = 1, 2, \ldots, 9$) has an $F$-distribution with $p - 1 = 8$ and $n - 9$ degrees of freedom. While $F_{(8,n-9,0.95)} < F_{(2,n-3,0.95)}$, it actually tends to become harder to reject the null hypothesis with more groups because we divide the numerator of $F$ in Equation (4.41) by $p - 1$, so our statistic is smaller. In other words, the power of our test is limited by the number of groups, as the degrees of freedom are expended on relatively rare genotypes. This well-known degrees-of-freedom problem is described for the analysis of haplotype–trait association studies in Chapman et al. (2003), Clayton et al. (2004), Tzeng et al. (2006) and Foulkes et al. (2008).

Several methods for addressing the degrees-of-freedom challenge have been described. For example, Chapman et al. (2003) propose an elegant approach involving the use of information on LD structure and selecting a subset of markers for analysis to minimize the degrees of freedom. Tzeng et al. (2006) propose a combination of dimension reduction, through the application of a clustering algorithm, and regression modeling based on homogenous haplotype groups. In this text, we present a modeling framework that addresses this challenge and can be seen as a simple extension of the ANOVA model of Equation (4.39). Further details of this approach are given in Goeman et al. (2004) for gene expression data and Foulkes et al. (2005) for SNP data.

Consider again the model given by Equation (4.39). We now make the additional assumption that the genotype group effects, $\alpha_i$, are random, arising from a specified distribution. Specifically, we assume

$$\alpha_i \sim N(0, \sigma_\alpha^2) \qquad (4.43)$$

where the variance parameter, $\sigma_\alpha^2$, is unknown and the $\alpha_i$ are independent of one another and independent of $\epsilon_{ij}$. This model is commonly referred to as a *random effects model*, or more generally a *mixed effects model* when additional covariates are included as independent variables. The mixed modeling framework is a well-established analytic approach for handling correlated observations arising from repeated measures on a single individual or clustering.

Clustering arises for a variety of reasons, including for example the inclusion within our study of multiple individuals within the same family unit. Several texts describe the mixed model in detail, including its application and methods for estimation and inference. See for example Diggle *et al.* (1994), Pinheiro and Bates (2000), Fitzmaurice *et al.* (2004) and Demidenko (2004).

In place of $H_0$ given by Equation (4.40), based on the mixed model, we can test the composite null hypothesis given by

$$H_0 : \sigma_\alpha^2 = 0 \tag{4.44}$$

In other words, the null hypothesis is that there is no variability in the effects of genotypes on the trait. A likelihood ratio test can be applied to test this hypothesis, and the resulting test statistic has a $\chi_0^2 + \chi_1^2 = \frac{1}{2}\chi_1^2$ distribution. We have a mixture distribution since we are testing a variance parameter at a boundary. Notably, this test has a single degree of freedom, regardless of the number of genotypes under consideration. This approach is straightforward to apply using the `lme()` and `nlme()` of the `nlme` package in R and is left as an exercise for the reader.

Predicted values of the random effects, $\alpha_i$, given by the empirical Bayes estimates (posterior means), and corresponding prediction intervals, provide additional information on the specific genotypes that may be driving the variability in effects and may be a useful exploratory tool. Note that in the mixed model we are assuming a prior distribution on genotype effects, which is typical of Bayesian inference. The application of a fully Bayesian approach to data arising from a GWAS that similarly begins by assuming a distribution on the SNP effects is described in Schumacher and Kraft (2007). This approach has its roots in Bayesian variable selection methods, as described in Section 7.4.

# Problems

**4.1.** Define and contrast the following terms: (1) family-wise error, (2) type-1 and type-2 errors, (3) strong and weak controls, and (4) complete and partial nulls.

**4.2.** Use the Bonferroni and the Benjamini and Hochberg corrections to test for an association between high-density lipoprotein cholesterol, represented by HDL_C, and all SNPs within the `actn3` gene based on the FAMuSS data. Code the SNPs as binary indicators for the presence of at least one variant allele. Compare and contrast these results with each other and with an analysis that is not adjusted for multiple comparisons.

**4.3.** Using the FAMuSS data and the Tukey adjustment, determine if there is a significant association between total cholesterol, measured by CHOL, and the `resistin_c180g` SNP. Perform the analysis overall and stratified by gender. Compare and contrast your findings.

**4.4.** Calculate the $q$-values corresponding to the tests described in Problem 4.2.

**4.5.** Write an R script to implement Scheffe's adjustment for multiple testing. Demonstrate it with an application to the Virco data.

**4.6.** Apply the free step-down resampling approach to adjust for multiple testing using the FAMuSS data to determine whether there is an association between the presence of at least one variant allele in each of the four SNPs within the `actn3` gene and the binary trait defined as an indicator for being in the top sample quartile of change in non-dominant arm muscle strength, given by `NDRM.CH`.

**4.7.** Determine the effective number of tests if we were to test for an association between each of the four `AKT1` SNPs and a trait using the HGDP data.

**4.8.** Based on the FAMuSS data, use a mixed effects model to test for an overall effect of the `akt2` gene on percentage change in non-dominant muscle strength, as measured by `NDRM.CH`, adjusted for `Race`, `Gender` and `Age`.

# 5

# Methods for Unobservable Phase

One of the primary analytic challenges in population-based genetic investigations of unrelated individuals is the unobservable nature of allelic phase. We introduced this concept briefly in Section 2.3.2, and here we elaborate on the statistical challenges and analytic techniques for characterizing haplotype associations in the context of unknown phase. Recall that *haplotypic phase* refers to the specific alignment of alleles on a single homologous chromosome and is generally not observable in the context of population-based investigations of unrelated individuals. Since the SNPs under study are often markers for the true disease-causing variant, haplotypes may capture more variability in the disease trait than genotype alone. Several statistical approaches to inferring haplotypic phase have been proposed. The goals of these methods are generally twofold. On the one hand, interest lies in estimating population-level haplotype frequencies; that is, the prevalence of specific haplotypes in the general population. Investigators are also interested in making inference about the association between haplotypes and a trait. This chapter addresses both aims and is divided into two sections. The first (Section 5.1) focuses on methods for estimation of haplotype frequencies that do not involve knowledge about a trait or disease phenotype. The second (Section 5.2) focuses on methods that involve both estimation of haplotype frequencies and testing for association between these haplotypes and a measured trait.

In the first section, two methods are described: (1) an expectation-maximization (EM) algorithm and (2) a Bayesian haplotype reconstruction approach. Both approaches draw solely on genotype information to arrive at haplotype estimates and do not incorporate knowledge about the trait under investigation. Further details on these approaches can be found in Excoffier and Slatkin (1995), Hawley and Kidd (1995), Long *et al.* (1995), Stephens and Donnelly (2000), Stephens *et al.* (2001) and Stephens and Donnelly (2003). Both methods can be used to infer individual-level haplotypes and in turn make inference about haplotype–trait associations; however, this must proceed with careful consideration of the additional variability introduced due to the uncertainty in the estimated data. This is described in Section 5.2.

A.S. Foulkes, *Applied Statistical Genetics with R: For Population-based Association* 129
*Studies*, Use R, DOI: 10.1007/978-0-387-89554-3_5,
© Springer Science+Business Media LLC 2009

Section 5.2 also considers a fully likelihood-based approach that additionally incorporates information about the trait of interest in the haplotype estimation procedure. Details of this approach can be found in Lake *et al.* (2003) and Lin and Zeng (2006). This latter approach provides a natural framework for formally testing haplotype–trait associations.

## 5.1 Haplotype estimation

In this section we discuss two methods for estimating individual haplotypes and population-level frequencies. The first is an EM approach that sets out to estimate haplotype frequencies. In turn, these estimates can be used to infer unknown haplotypes for the individuals in our sample. The second approach we describe is a Bayesian approach that focuses on reconstructing unknown haplotypes. In turn, the reconstructed data can be used to estimate population-level haplotype frequencies. More detailed mathematical derivations require knowledge of elementary calculus and can be found in the supplemental notes at the end of this chapter. Here emphasis is placed on some general intuition behind the approaches and the practical tools for their implementation. Methods for using the results of these analyses for the study of haplotype–trait associations are described in Section 5.2.

### 5.1.1 An expectation-maximization algorithm

The expectation-maximization (EM) algorithm is a natural approach to estimating population-level parameters in the context of missing data. Details on the original algorithm and further intuition behind its inception are given in Dempster *et al.* (1977). In order to understand intuitively how this algorithm works, recall that a maximum likelihood estimate (MLE) is an estimate of a population-level parameter, $\theta$, that is derived by maximizing a function of the data, $\mathbf{X} = (x_1, \ldots, x_n)$. Specifically, the MLE is the value of the parameter that maximizes the *likelihood* of the data, given by

$$L(\theta|\mathbf{X}) = \prod_{i=1}^{n} Pr(x_i|\theta) \qquad (5.1)$$

where $Pr(x_i|\theta)$ is the probability density function of $x_i$. Since taking the logarithm of a function is an order-preserving transformation, maximizing this function is equivalent to maximizing the logarithm of this function, which tends to be easier analytically. In missing data settings, the data $X$ are not fully observed and so the likelihood of Equation (5.1) cannot be calculated. The set of data that includes both the observed data, denoted $\mathbf{X}^{obs}$, and the missing data is commonly referred to as the *complete* data and is denoted $\mathbf{X}^c$.

The EM approach involves first taking the expectation of the complete data log likelihood conditional on the observed data and the current parameter estimate. This is called the *E-step* and amounts to determining the most likely value of the likelihood for the complete data given what we have observed and the current estimate $\widehat{\theta}^{(t)}$. Formally, we calculate

$$E\left(\log L(\theta|\mathbf{X}^c)|\mathbf{X}^{obs}, \widehat{\theta}^{(t)}\right) \qquad (5.2)$$

where $E(\cdot)$ is the expectation and is defined in Section 2.1.1. The second step of the EM algorithm, called the *M-step*, is to maximize Equation (5.2) with respect to the parameter $\theta$. This yields a new estimate, which we denote $\widehat{\theta}^{(t+1)}$. Finally we iterate between the E-step and the M-step until a convergence criterion is met to arrive at an MLE of $\theta$.

Details of the algorithm above for the genetic association setting are given in the supplemental notes at the end of this chapter. In our setting, interest lies in estimating the population-level haplotype frequencies. That is, $\theta$ is a vector with elements corresponding to the probability of each haplotype. The observed data are the genotypes for all individuals in our sample, and the missing data are the corresponding haplotype pairs. In other words, the observed data are the two nucleotides at each site, and the missing data are the specific alignment of these nucleotides on each of the two homologous chromosomes. The complete data are thus comprised of both the observed genotype information and the haplotype pairs. Since knowledge of haplotype pairs will inform us about genotypes, we more simply express the complete data as the haplotype pairs.

The EM algorithm proceeds by first writing out the complete data log likelihood for the haplotypes and then taking the conditional expectation of this function, conditional on the observed genotypes. This expectation is a weighted sum over all haplotype pairs that are consistent with the observed genotypes for a given individual and is written formally as

$$E\left(\log L(\theta|\mathbf{X}^c)|\mathbf{X}^{obs}, \widehat{\theta}^{(t)}\right) = E\left(\log L(\theta|H_1, \ldots, H_n) \mid G_1, \ldots, G_n, \widehat{\theta}^{(t)}\right)$$

$$= \sum_{i=1}^{n} \sum_{H_i \in \mathcal{S}(G_i)} \widehat{p}_{H_i}^{(t)} \log Pr(H_i|\theta)$$

$$(5.3)$$

where $G_i$ is the observed genotype for individual $i$, $H_i$ is a corresponding haplotype pair and $\mathcal{S}(G_i)$ is the set of all haplotype pairs that are consistent with the observed genotype. For example, suppose the observed genotype across two sites for individual $i$ is given by $G_i = AaBb$. The set of all haplotype pairs that are consistent with this genotype is given by $\mathcal{S}(G_i) = \{(AB, ab), (Ab, aB)\}$. The weight, given by $\widehat{p}_{H_i}^{(t)}$ for individual $i$, is the estimated probability of $H_i$ given the genotype for this individual and $\widehat{\theta}^{(t)}$. Formally, we refer to these weights as *posterior probabilities* of the corresponding haplotype pair given the observed genotype and current parameter estimates and write

$$\widehat{p}_{H_i}^{(t)} = Pr(H_i | G_i, \widehat{\theta}^{(t)}) = \frac{Pr(H_i, G_i | \widehat{\theta}^{(t)})}{Pr(G_i | \widehat{\theta}^{(t)})} = \frac{Pr(H_i | \widehat{\theta}^{(t)})}{\sum_{H_i \in S(G_i)} Pr(H_i | \widehat{\theta}^{(t)})} \quad (5.4)$$

Intuitively, the idea behind the E-step is that we are averaging over all possible resolutions of the missing data in a manner that takes into account the current parameter estimates. Consider for example the simple case above in which an individual's genotype is given by $Aa, Bb$. Application of the E-step will give more weight to the haplotype pair that has a higher estimated frequency. That is, suppose the haplotypes $AB$ and $ab$ are relatively common while the haplotypes $Ab$ and $aB$ appear rare, based on our current parameter estimates. In this case, we want to lend additional weight to the $(AB, ab)$ pair since it is more likely to be the true haplotype for this individual. The posterior probabilities capture this information. During the M-step, the expectation of Equation (5.3) is maximized to arrive at updated parameter estimates and the process is repeated. Importantly, this approach assumes HWE and thus should be applied within racial and ethnic strata within which there is no evidence of a departure from this assumption. Implementation of this procedure is straightforward, as illustrated in the following example. Detailed derivations are provided in the supplemental material at the end of this chapter.

*Example 5.1 (EM approach to haplotype frequency estimation).* In this example, we estimate the population-level frequencies of haplotypes within the actn3 gene for African Americans and Caucasians separately based on the FAMuSS data. We begin by calling the haplo.stats package and creating a genotype matrix. The genotype matrix has a pair of adjacent columns for each SNP such that each column corresponds to one of the two observed alleles at the corresponding site. The order of the columns is assumed to correspond to the order of the sites on the chromosome. Recall that we start with four SNPs within the actn3 gene and so the following code is required:

```
> install.packages("haplo.stats")
> library(haplo.stats)
> attach(fms)
> Geno <- cbind(substr(actn3_r577x,1,1), substr(actn3_r577x,2,2),
+        substr(actn3_rs540874,1,1), substr(actn3_rs540874,2,2),
+        substr(actn3_rs1815739,1,1), substr(actn3_rs1815739,2,2),
+        substr(actn3_1671064,1,1), substr(actn3_1671064,2,2))
> SNPnames <- c("actn3_r577x", "actn3_rs540874", "actn3_rs1815739",
+        "actn3_1671064")
```

We then subset African Americans and Caucasians and apply the haplo.em() function to each group. This function applies a modified version of the EM approach described above, in which sets of loci are progressively included and in turn haplotype pairs with small estimated probabilities are excluded. The haplo.em.control() function is used within the haplo.em() function call to specify the minimum posterior probability of a haplotype pair. Pairs that

have an estimated frequency lower than this threshold will be removed from
the list of possible pairs.

```
> Geno.C <- Geno[Race=="Caucasian" & !is.na(Race),]
> HaploEM <- haplo.em(Geno.C, locus.label=SNPnames,
+        control=haplo.em.control(min.posterior=1e-4))
> HaploEM
# Note that the results may differ slightly each run since different
# starting values are used
```

```
==========================================================================
                              Haplotypes
==========================================================================

    actn3_r577x actn3_rs540874 actn3_rs1815739 actn3_1671064 hap.freq
1        C             A              C              G      0.00261
2        C             A              T              A      0.00934
3        C             A              T              G      0.01354
4        C             G              C              A      0.47294
5        C             G              C              G      0.01059
6        T             A              C              A      0.00065
7        T             A              T              G      0.39891
8        T             G              C              A      0.08557
9        T             G              T              A      0.00065
10       T             G              T              G      0.00520
==========================================================================
                                Details
==========================================================================
lnlike =  -1285.406
lr stat for no LD =  2780.769 , df =  5 , p-val =  0
```

```
> Geno.AA <- Geno[Race=="African Am" & !is.na(Race),]
> HaploEM2 <- haplo.em(Geno.AA, locus.label=SNPnames,
+        control=haplo.em.control(min.posterior=1e-4))
> HaploEM2
```

```
==========================================================================
                              Haplotypes
==========================================================================
    actn3_r577x actn3_rs540874 actn3_rs1815739 actn3_1671064 hap.freq
1        C             A              C              A      0.01140
2        C             A              C              G      0.08130
3        C             A              T              G      0.03764
4        C             G              C              A      0.57762
5        C             G              C              G      0.01156
6        T             A              C              A      0.00032
7        T             A              T              G      0.17166
8        T             G              C              A      0.10833
9        T             G              C              G      0.00016
==========================================================================
                                Details
```

```
=========================================================================
lnlike =  -84.97891
lr stat for no LD =   119.7087 , df =  4 , p-val =  0
```

The column entitled `hap.freq` is the estimated population-level haplo-type frequency. Note that the row numbers in the two outputs above do not necessarily correspond to the same haplotypes. Based on this output, we can see that the most prevalent haplotype is the same in African Americans and Caucasians and is given by $h_4 = $ `CGCA`. The estimated prevalence of this haplotype is higher for African Americans ($\widehat{\theta}_4 = 0.58$) than for Caucasians ($\widehat{\theta}_4 = 0.47$). On the other hand, the estimated prevalence of the $h_7 = $ `TATG` haplotype is markedly lower in African Americans ($\widehat{\theta}_7 = 0.17$) than in Caucasians ($\widehat{\theta}_7 = 0.40$). $\qquad\qquad\qquad\qquad\qquad\square$

Based on the estimated population-level haplotype probabilities, we can calculate the probabilities of each possible haplotype pair for an observation in our sample. Consider again the simple example in which an individual presents with the genotype $Aa$ and $Bb$ across two SNPs. This individual's haplotype pair is ambiguous, though we know it is either $H_1 = (AB, ab)$ or $H_2 = (Ab, aB)$. Now suppose the haplotype frequencies are $\theta_1, \theta_2, \theta_3$ and $\theta_4$ for haplotypes $AB, Ab, aB$ and $ab$, respectively. Assuming independence, we know that the posterior probability of $H_1$ is $p_1 = 2\theta_1\theta_4$, while the probability of $H_2$ is $p_2 = 2\theta_2\theta_3$ given the observed genotype. If additional haplotype pairs are present in our population, we must additionally divide each of these probabilities by the sum $p_1 + p_2$ since we are conditioning on one of the two haplotype pairs for this individual. A demonstration of how we can calculate these posterior probabilities is given in the following example.

*Example 5.2 (Calculating posterior haplotype probabilities).* In this example, we illustrate how to determine the posterior probability of each haplotype pair that is consistent with the observed genotype for an individual. Let us return to Example 5.1 and recall that `HapoEM` is the result of applying the `haplo.em()` function to the SNPs within the `actn3` gene on the Caucasian subgroup within the FAMuSS study. The associated object `HaploEM$nreps` is a vector of length equal to the number of individuals in our sample with elements equal to the number of haplotype pairs that are consistent with the observed genotype. For example, consider the first five elements of this vector, given by

```
> HaploEM$nreps[1:5]

indx.subj
1 2 3 4 5
1 2 2 2 1
```

This tells us that there is one haplotype pair consistent with the observed genotype for the first and fifth individuals and two pairs that are consistent with each of the observed genotypes for the second, third and fourth individuals. The corresponding potential haplotypes for these five individuals are given by the associated vectors `HaploEM$hap1code` and `HaploEM$hap2code` as shown below, where the coding corresponds to the numbering system we saw for Caucasians in Example 5.1. The `indx.subj` vector tells us the corresponding record number and contains the sequence of numbers from 1 to the number of observations in our sample, with each element of the sequence repeated according to the value in `HaploEM$nreps`.

```
> HaploEM$indx.subj[1:8]

[1] 1 2 2 3 3 4 4 5

> HaploEM$hap1code[1:8]

[1] 4 8 7 3 7 8 4 4

> HaploEM$hap2code[1:8]

[1] 4 3 4 8 4 3 7 4
```

Based on this output, we see that the first and fifth individuals have the haplotype pair $(4,4)$, while the second, third and fourth individuals are all ambiguous between $(3,8)$ and $(4,7)$. The posterior probabilities associated with these pairs are given by

```
> HaploEM$post[1:8]

[1] 1.000000000 0.006102808 0.993897192 0.006102808 0.993897192
[6] 0.006102808 0.993897192 1.000000000
```

Notably, the sum of these probabilities within any single individual is equal to 1.

We can also calculate these probabilities directly based on the estimated haplotype frequencies. To see this, first note that the ten haplotype probabilities given in the first table of output in Example 5.1 are contained in the vector

```
> HapProb <- HaploEM$hap.prob
> HapProb

 [1] 0.0026138447 0.0093400121 0.0135382727 0.4729357032 0.0105890282
 [6] 0.0006518550 0.3989126969 0.0855667219 0.0006548104 0.0051970549
```

Now consider one of our individuals who is ambiguous between the pairs $(3,8)$ and $(4,7)$. Assuming independence, which was already assumed in the estimation procedure, estimated probabilities of each of these pairs are given respectively by

```
> p1 <- 2*prod(HapProb[c(3,8)])
> p2 <- 2*prod(HapProb[c(4,7)])
> p1 / (p1+p2)
```

```
[1] 0.006102807

> p2 / (p1+p2)

[1] 0.9938972
```

As expected, these values are equivalent to the probabilities given in the second and third elements of `HaploEM$post`.    □

Investigators often fill in unknown haplotypes by assigning each individual the haplotype pair with the highest corresponding posterior probability and then treating these as known in subsequent analysis. We caution the reader against proceeding in this manner since valuable information on the uncertainty in the assignment is lost. Instead, methods described in Section 5.2 can be applied if the ultimate goal is to characterize haplotype–trait association.

Finally, we note that testing hypotheses involving haplotype frequencies within the EM context requires consideration of the uncertainty in the estimation procedure. For example, suppose we are interested in testing the null hypothesis that the $h_4 = CGCA$ haplotype frequencies are equal for Caucasians and African Americans. This requires knowledge about the unknown variance/covariance matrix of the estimates. Derivation of this matrix can be obtained by inverting the observed information matrix and using Louis' method for the EM framework (Louis, 1982). Alternatively, the observed information matrix can be approximated with the empirical observed information (Meilijson, 1989; McLachlan and Krishnan, 1997). The details of this derivation are beyond the scope of this textbook, though an example of applying this approach is given below.

*Example 5.3 (Testing hypotheses about haplotype frequencies within the EM framework).* In this example, we test the null hypothesis that the frequency of haplotype $h_4$ defined in Example 5.1 is the same in African Americans and Caucasians. We do this by constructing a confidence interval around the difference in the two frequencies and checking whether it covers 0. First we calculate the difference in the estimated frequencies. In order to calculate the standard error of each frequency, we use the function `HapFreqSE()` defined in the supplemental R scripts at the end of this chapter. We then combine the results to get an estimate of the standard error of this difference. Finally, a 95% confidence interval is calculated based on a normal probability distribution:

```
> FreqDiff <- HaploEM2$hap.prob[4] - HaploEM$hap.prob[4]
> s1 <- HapFreqSE(HaploEM)[4]
> s2 <- HapFreqSE(HaploEM2)[4]
> SE <- sqrt(s1^2 + s2^2)
> CI <- c(FreqDiff - 1.96*SE, FreqDiff + 1.96*SE)
> CI

[1] -0.003395297  0.212772537
```

Since this interval covers 0, there is not enough evidence to suggest a difference in the frequency of $h_4$ between Caucasians and African Americans based on our sample and a two-sided 0.05 level test. □

## 5.1.2 Bayesian haplotype reconstruction

Several alternative Bayesian approaches have been described for analyzing data in the context of missing haplotypic phase (Stephens *et al.*, 2001; Niu *et al.*, 2002; Lin *et al.*, 2002; Stephens and Donnelly, 2003). Here we focus on the haplotype reconstruction approach of Stephens *et al.* (2001) and related extensions. Similar to the EM approach described above, this method allows for estimation of population-level haplotype frequencies in the context of data for which allelic phase is potentially unobservable. The primary aim, however, is the *reconstruction* of individual-level haplotype pairs. That is, the approach we present sets out to assign each individual the most likely haplotype pair. Estimation of haplotype frequencies then follows, assuming these haplotypes are the true haplotypes. In this section, we focus on the approach of Stephens *et al.* (2001) with extensions described in Stephens and Donnelly (2003). This method can be implemented using the PHASE and fastPHASE software, though a comparable R package is not yet available. We begin by providing a brief discussion of Bayesian inference and a computational technique called Gibbs sampling. The reader is referred to an intermediate text for additional discussion and examples of Bayesian methods (Gelman *et al.*, 2004; Givens and Hoeting, 2005).

The general idea behind Bayesian methods is that we can make inference about our parameter based on its conditional distribution given the data. Let the parameter of interest be denoted $\theta$ and the data be given by $\mathbf{X}$. The conditional distribution of $\theta$ given $\mathbf{X}$ is denoted $\pi(\theta|\mathbf{X})$ and is commonly referred to as the *posterior density* of $\theta$. This distribution depends on three quantities: (1) the prior distribution of $\theta$, given by $\pi(\theta)$; (2) the likelihood of the data, given by $L(\theta|\mathbf{X}) = f(\mathbf{X}|\theta)$; and (3) a constant, denoted $c$ and written formally as $c = 1/\left(\int_\theta \pi(\theta)L(\theta|\mathbf{X})d\theta\right)$. Recall that, in the more traditional *frequentist* setting, estimation is based instead on the likelihood $L(\theta|\mathbf{X}) = f(\mathbf{X}|\theta)$. The relationship between the posterior density and each of the three quantities listed is a result of Bayes' rule, given by

$$\pi(\theta|\mathbf{X}) = \frac{\pi(\theta;\mathbf{X})}{f(\mathbf{X})} = \frac{f(\mathbf{X}|\theta)\pi(\theta)}{\int_\theta \pi(\theta)f(\mathbf{X}|\theta)d\theta} \tag{5.5}$$

Equivalently, we can write

$$\pi(\theta|\mathbf{X}) = cL(\theta|\mathbf{X})\pi(\theta) \tag{5.6}$$

since the integral in the denominator of Equation (5.5) is a constant that does not depend on $\theta$, and $f(\mathbf{X}|\theta)$ is the likelihood of $\theta$ given the observed data,

**X**. In practice, exact calculation of this posterior distribution is not tenable and computational techniques for approximating it are needed. Markov chain Monte Carlo (MCMC) methods are an approach to generating an approximate sample from a distribution, and one well-described example of an MCMC approach is the Gibbs sampler.

Here we present a brief description of the Gibbs sampler in a general context and then discuss its application to the genetics setting. Suppose the population parameters are $\boldsymbol{\theta} = (\theta_1, \ldots, \theta_K)$ and that we are interested in the joint posterior density, $\pi(\boldsymbol{\theta}|\mathbf{X})$, which we cannot obtain analytically. Further suppose $\pi(\theta_k|\theta_{-k}, \mathbf{X})$ is the marginal distribution of the single parameter $\theta_k$ conditional on current values of all other parameters, $\theta_1, \ldots, \theta_{k-1}, \theta_{k+1}, \ldots, \theta_K$. A Gibbs sampler provides us with a sample of data from the posterior density, $\pi(\boldsymbol{\theta}|\mathbf{X})$, based on sampling from these marginal distributions, $\pi(\theta_k|\theta_{-k}, \mathbf{X})$. It proceeds as follows, where we begin by letting $t = 0$ and defining initial values for $\theta_1^{(0)}, \ldots, \theta_K^{(0)}$.

---

**ALGORITHM 5.1: GIBBS SAMPLING**
1. Sample:

- $\theta_1^{(t+1)}|\theta_{-1}, \mathbf{X} \sim \pi(\theta_1|\theta_2^{(t)}, \ldots, \theta_K^{(t)}, \mathbf{X})$

- $\theta_2^{(t+1)}|\theta_{-2}, \mathbf{X} \sim \pi(\theta_2|\theta_1^{(t+1)}, \theta_3^{(t)}, \ldots, \theta_K^{(t)}, \mathbf{X})$

- $\theta_3^{(t+1)}|\theta_{-3}, \mathbf{X} \sim \pi(\theta_3|\theta_1^{(t+1)}, \theta_2^{(t+1)}, \theta_4^{(t)}, \ldots, \theta_K^{(t)}, \mathbf{X})$

  $\vdots$

- $\theta_K^{(t+1)}|\theta_{-K}, \mathbf{X} \sim \pi(\theta_K|\theta_1^{(t+1)}, \ldots, \theta_{K-1}^{(t+1)}, \mathbf{X})$

2. Let $t = t + 1$ and repeat step (1) $M$ times for $M$ large.

---

We see that $\theta^{(t)}$ is the value of $\theta$ at the $t$th iteration and is sampled conditional on current estimates of the remaining parameters. Algorithm 5.1 results in what is termed a *Markov chain* since each parameter value $\theta_k^{(t+1)}$ depends only on the previous value $\theta_k^{(t)}$ in the series. Furthermore, under certain regularity conditions, it can be shown that the stationary distribution of this chain, as the number of iterations $s \to \infty$, is equal to $\pi(\theta|\mathbf{X})$. Thus, the final observation from this chain represents a sample point from the posterior distribution, $\pi(\theta|\mathbf{X})$. In practice, after convergence is met, we continue sampling using the same iterative algorithm to obtain a sample, $\theta^1, \ldots, \theta^S$, from the posterior distribution. Taking the mean across these $S$ samples will give a consistent estimate of $\theta$.

Now let us consider the genetics setting. Let $H_i$ represent the potentially unobservable haplotype and $G_i$ represent the observed genotype for individual

$i$, where $i = 1, \ldots, n$. We are interested in the posterior probability of haplotypes given the observed genotype data across individuals, denoted $\pi(\mathbf{H}|\mathbf{G})$. This probability relies on a prior distribution for the haplotypes, given by $\pi(\mathbf{H})$, and the likelihood of the data, given by $L(\mathbf{H}|\mathbf{G})$. Formally, we write

$$\pi(\mathbf{H}|\mathbf{G}) = cL(\mathbf{H}|\mathbf{G})\pi(\mathbf{H}) \tag{5.7}$$

where again $c$ is a constant. Application of a Gibbs sampler to reconstruct haplotypes proceeds as follows. Again the idea is to generate data from the posterior distribution and then find the mode of this distribution, which will give a consistent estimate of the haplotype pair for an individual. The algorithm is similar to the one described above, except that the parameters, given by $\boldsymbol{\theta}$, are replaced with haplotype pairs. Specifically, in its original formulation, the algorithm is given as follows, where we again begin by letting $t = 0$ and defining initial values $H_1^{(0)}, \ldots, H_{n^*}^{(0)}$, where $n^*$ is the number of ambiguous individuals in our sample:

---

ALGORITHM 5.2: BAYESIAN HAPLOTYPE RECONSTRUCTION

1. Sample:
   - $H_1^{(t+1)}|\mathbf{G}, \mathbf{H}_{-1} \sim \pi(H_1|H_2^{(t)}, \ldots, H_{n^*}^{(t)}, \mathbf{G})$

   - $H_2^{(t+1)}|\mathbf{G}, \mathbf{H}_{-2} \sim \pi(H_2|H_1^{(t+1)}, H_3^{(t)}, \ldots, H_{n^*}^{(t)}, \mathbf{G})$

   - $H_3^{(t+1)}|\mathbf{G}, \mathbf{H}_{-3} \sim \pi(H_3|H_1^{(t+1)}, H_2^{(t+1)}, H_4^{(t)}, \ldots, H_{n^*}^{(t)}, \mathbf{G})$

   $\vdots$

   - $H_{n^*}^{(t+1)}|\mathbf{G}, \mathbf{H}_{-n^*} \sim \pi(H_{n^*}|H_1^{(t+1)}, \ldots, H_{n^*-1}^{(t+1)}, \mathbf{G})$

2. Let $t = t + 1$, and repeat step (1) $M$ times for $M$ large.

---

Repeating Algorithm 5.2 a large number of times again results in a Markov chain and leads to consistent estimates of the unknown haplotypes. In order to implement this algorithm, we need to make some assumptions about the distribution $\pi(H_i|\mathbf{G}, \mathbf{H}_{-i})$ from which we sample. A cogent summary with implications for each underlying model is provided in Stephens and Donnelly (2003). The coalescence model given by Stephens and Donnelly (2000) has gained popularity due to its relatively strong performance and interpretability. Similar to the EM framework, this approach assumes random mating so that the probability of a pair of haplotypes is given by the product of each of the two haplotype probabilities. Further refinements of this algorithm, which improve overall performance, have been described by Stephens et al. (2001), Stephens and Donnelly (2003) and Scheet and Stephens (2006). These extensions are straightforward to implement using the publicly available PHASE and fastPHASE software. Once each individual's unknown haplotype pair is reconstructed, we can estimate population-level haplotype frequencies by simply tabulating the data under the assumption that the individual haplotype pairs are known.

One advantage of PHASE is in its handling of ambiguous individuals whose possible haplotype pairs are not any of the observed haplotypes. The PHASE approach assigns these individuals pairs of haplotypes that are similar to observed haplotypes with high population-level frequencies. That is, PHASE uses information on the frequencies of observed haplotypes to inform us about the likelihood of similar haplotypes within phase-ambiguous individuals. This is in contrast to the EM approach described in Section 5.1.1, which instead will give haplotype pairs equal posterior probabilities if they are not observed. In practice, the two approaches will tend to differ if the number of SNPs under study is large and/or several alleles are present at a given SNP site.

## 5.2 Estimating and testing for haplotype–trait association

In the previous section, focus was on methods for haplotype estimation that are based solely on genetic sequence information. More generally, information on a measured trait is also available, and interest lies in detecting association between haplotypes and this trait. In this case, haplotype reconstruction may be an intermediary step in a larger hypothesis-driven analysis. Notably, information on a trait can also help to inform us about the true underlying haplotypes, as described in Section 5.2.2. In this section, we describe two general approaches to estimating and testing for association between a haplotype and a trait in the context of population-based data in which allelic phase is potentially ambiguous. The first set of approaches, given in Section 5.2.1, are two-staged approaches that involve first reconstructing haplotypes using one of the methods described in Section 5.1 and then fitting a standard generalized linear model. The second approach, discussed in Section 5.2.2, is fully likelihood-based and uses an EM-type algorithm that incorporates both the genotype and trait in simultaneous estimation of population-level haplotype frequencies and haplotype–trait associations.

### 5.2.1 Two-stage approaches

The methods described in Section 5.1 provide us with reconstructed haplotypes based on the most likely pair given the observed genotype information. The most straightforward approach to characterizing haplotype–trait associations is to assume simply these reconstructed haplotypes are the true haplotypes and fit a model of association. In the statistics literature, this is commonly referred to as *single imputation* since we are completing the data a single time. Unfortunately, while simple, this approach can lead to erroneous inference about our haplotype–trait associations. For a full discussion of the pitfalls of this approach, see Lin and Huang (2007) and the associated commentary by Kraft and Stram (2007). Here we describe two alternatives to single imputation that both draw on the results of the analyses presented in Section 5.1.

*Haplotype trend regression*

Haplotype trend regression (HTR) is an approach to association analysis that involves assigning to each ambiguous individual in our sample the *conditional expectation* of the number of copies of each potential haplotype. That is, in place of indicating the presence of $n_h = 0, 1$ or 2 copies of haplotype $h$, as we would do in the observed data setting, we instead assign a value of $E(n_h|G)$. Consider for example the simple situation in which an individual $i$ is ambiguous between the haplotype pairs $H_1 = (h_1, h_4)$ and $H_2 = (h_2, h_3)$ with posterior probabilities of $p_1$ and $p_2$, respectively. In this case, we define the variables $x_{i1}, \ldots, x_{i4}$, where $x_{ij}$ is the value corresponding to the $j$th haplotype for individual $i$, and set $x_{i1} = x_{i4} = p_1$ and $x_{i2} = x_{i3} = p_2$. Finally, we test for both overall and specific haplotype–trait associations by fitting a linear model that includes these $x_{ij}$'s as predictor variables and conduct the usual $F$-test. We demonstrate this in the following example. Further details of this approach can be found in Zaykin *et al.* (2001).

*Example 5.4 (Application of haplotype trend regression (HTR)).* Haplotype trend regression begins by creating a design matrix with elements equal to the conditional expectation for the number of copies of each haplotype. This is achieved using the function `HapDesign()` defined in the supplemental notes at the end of this chapter. Fitting a linear model using this new design matrix and conducting an overall $F$-test that compares this model to the reduced model with just an intercept is straightforward. In the following code, we use the percentage change in muscle strength for the non-dominant arm as the trait and the `HaploEM` object from Example 5.1:

```
> attach(fms)
> HapMat <- HapDesign(HaploEM)
> Trait <- NDRM.CH[Race=="Caucasian" & !is.na(Race)]
> mod1 <- (lm(Trait~HapMat))
> mod2 <- (lm(Trait~1))
> anova(mod2,mod1)

Analysis of Variance Table

Model 1: Trait ~ 1
Model 2: Trait ~ HapMat
  Res.Df    RSS Df Sum of Sq      F Pr(>F)
1    776 881666
2    764 868364 12     13303 0.9753 0.4708
```

Based on this output, we conclude that there is not sufficient evidence to support a haplotype–trait association. □

*Multiple imputation*

An alternative to the HTR approach is *multiple imputation (MI)*, a well-described approach for handling missing data that involves repeatedly filling in missing information and then making an appropriate adjustment in the inference procedure. This adjustment is important to account for the additional variability introduced by the uncertainty in the "realized" data values. A complete discussion of MI can be found in Little and Rubin (2002). Application of MI in the context of phase uncertainty is straightforward using the posterior probability estimates that we derived in Section 5.1.

Suppose $\mathbf{p}_i$ is a vector of estimated posterior probabilities for each haplotype pair $H_i \in \mathcal{S}(G_i)$ for individual $i$, where $\mathcal{S}(G_i) = \{H_{i1}, \ldots, H_{iK_i}\}$ is again the set of all haplotype pairs that are consistent with the observed genotype, $G_i$. An explicit formula for the posterior probabilities is given by $\widehat{p}_{H_i}(\theta)$ in Equation (5.4). If an individual's haplotype pair is fully determined given the observed genotype, then the size of $\mathcal{S}(G_i) = 1$ and $\mathbf{p}_i = 1$. Multiple imputation involves repeated sampling of the possible haplotype pairs for each individual based on the probabilities in $\mathbf{p}_i$. Specifically, we repeatedly assign $H_i^* = H_{ik}$ with probability $\widehat{p}_{ik}$ for $k = 1, \ldots, K_i$. In the simple case where $\mathbf{p}_i = 1$, $H_i^*$ is set equal to the observed haplotype pair. Performing this one time is called a single imputation, while repeating this sampling process multiple times is referred to as multiple imputation. Once an imputed dataset is generated, a model-fitting procedure can be applied and a corresponding effect estimate and test statistic can be recorded. These values can then be combined over the multiply imputed datasets.

To understand how this approach works in practice, consider the simple linear regression model described in Section 2.2.3 and given by the equation

$$y_i = \alpha + \beta x_i + \epsilon_i \qquad (5.8)$$

Here $y_i$ is a quantitative trait and we define $x_i = I(h_k \in H_i)$ as an indicator for whether haplotype $h_k$ is in the (potentially unobserved) diplotype $H_i$ for individual $i$, where $i = 1, \ldots, n$. Further suppose interest lies in testing the null hypothesis of no association between $x$ and $y$, given by $H_0 : \beta = 0$. As discussed in Section 2.2.3, a Wald statistic can be used to test this null hypothesis. In the missing data context, we instead calculate $\widehat{\beta}$ for each of multiple imputed datasets. That is, we complete the data by randomly selecting a haplotype pair for each individual with probabilities equal to the estimated posterior probabilities resulting from application of the EM approach. Based on these completed data, we fit the model of Equation (5.8) and arrive at a least squares estimate of $\beta$. This process is then repeated $D$ times, and we let the resulting estimates be denoted $\widehat{\beta}_1, \ldots, \widehat{\beta}_D$.

A combined estimate of $\beta$ over these $D$ imputations is then given by

$$\bar{\beta}_D = \frac{1}{D} \sum_{d=1}^{D} \widehat{\beta}_d \qquad (5.9)$$

The variance of this estimate must take into account the variability both within and between imputations. Suppose $\widehat{W}_d$ is the variance of $\widehat{\beta}$ for the $d$th completed dataset, where again $d = 1, \ldots, D$. The within-imputation variance, denoted $\bar{W}_D$, can be calculated as the average over the $D$ imputations, and the between-imputation variance, denoted $\bar{B}_D$, is given by a summary of the deviations of each estimated value of $\widehat{\beta}$ from the average over all imputations. The total variability that is associated with the parameter estimate, given by $T_D$, is a weighted sum of the within- and between-imputation variances. Formally, we write

$$\bar{W}_D = \frac{1}{D} \sum_{d=1}^{D} \widehat{W}_d \qquad (5.10)$$

$$\bar{B}_D = \frac{1}{D-1} \sum_{d=1}^{D} \left( \widehat{\beta}_d - \bar{\beta}_D \right)^2 \qquad (5.11)$$

$$T_D = \bar{W}_D + \frac{D+1}{D} \bar{B}_D \qquad (5.12)$$

Note that the total variability will be greater than the variability based on any single imputation. Furthermore, as the difference between estimates from one imputation to the next approaches 0, this total variability will approach the average of the within-imputation variances. A $t$-test statistic corresponding to the null hypothesis $H_0 : \beta = 0$ is now given by

$$\bar{\beta}_D / \sqrt{T_D} \sim t_v \qquad (5.13)$$

where the degrees of freedom, $v$, are equal to

$$v = (D-1) \left( 1 + \frac{1}{D+1} \frac{\bar{W}_D}{\bar{B}_D} \right)^2 \qquad (5.14)$$

A step-by-step summary of this MI approach is provided in Algorithm 5.3. Notably, the first step of this algorithm involves estimation of the posterior probabilities. This introduces additional variability that we are not accounting for in this procedure. Bootstrapping the entire algorithm would capture this additional variability. This involves resampling from our original data, with replacement, running through the algorithm and then repeating multiple times. An application of MI to estimation and testing of haplotype–trait association is given in the following example.

---

ALGORITHM 5.3: MULTIPLE IMPUTATION FOR HAPLOTYPE–TRAIT ASSO-
CIATION

1. Estimate posterior probabilities of each haplotype pair for each
   individual $i = 1, \ldots, n$ given the observed genotypes.

2. Sample a single diplotype for each individual according to this probability distribution.

3. Calculate an effect estimate and corresponding variance.

4. Repeat steps (2) and (3) $D$ times for $D$ large.

5. Determine and evaluate test statistic according to Equation (5.13).

---

*Example 5.5 (Multiple imputation for haplotype effect estimation and testing).*
In this example, we again consider the question of whether there is an association between the actn3 gene and the percentage change in the non-dominant
arm muscle strength before and after exercise training. Specifically, we consider whether the presence of at least one copy of the $h_8 = $ TGCA haplotype
within Caucasians is associated with NDRM.CH using the FAMuSS sample. The
following for() loop samples from the potential haplotypes for each individual
according to the estimated posterior probabilities. This is repeated $D = 1000$
times, and the estimated effects of the $h_9$ haplotype and corresponding standard errors are recorded in the vectors Est and SE:

```
> attach(fms)
> Nobs <- sum(Race=="Caucasian", na.rm=T)
> Nhap <- length(HaploEM$hap.prob)
> D <- 1000
> Est <- rep(0,D)
> SE <- rep(0,D)
> for (nimput in 1:D){
+        Xmat <- matrix(data=0,nrow=Nobs,ncol=Nhap)
+        for (i in 1:Nobs){
+              IDSeq <- seq(1:sum(HaploEM$nreps))[HaploEM$indx.subj==i]
+              if (length(IDSeq)>1){Samp <- sample(IDSeq,size=1,
+                      prob=HaploEM$post[IDSeq])}
+              if (length(IDSeq)==1){Samp <- IDSeq}
+              Xmat[i,HaploEM$hap1code[Samp]] <-1
+              Xmat[i,HaploEM$hap2code[Samp]] <-1
+              }
+        h8 <- Xmat[,8]>=1
+        Est[nimput] <- summary(lm(Trait~h9))$coefficients[2,1]
+        SE[nimput] <- summary(lm(Trait~h9))$coefficients[2,2]
+        }
# (this can take several minutes to run)
```

Based on these results, we can calculate the average estimate over all the imputations (`MeanEst`), the within-imputation variance (`Wd`), the between-imputation variance (`Bd`) and the total variance (`Td`) as follows:

```
> MeanEst <- mean(Est)
> Wd <- mean(SE^2)
> Bd <- (1/(D-1))*sum((Est-MeanEst)^2)
> Td <- Wd + ((D+1)/D)*Bd
```

Finally, we calculate the degrees of freedom and $p$-value corresponding to a test of no association between the presence of at least one copy of $h_8$ and the percentage change in muscle strength of the non-dominant arm. Here we use the `pt()` function, which returns the lower tail probability corresponding to the quantile and degrees of freedom given as its argument.

```
> nu <- D-1*(1 + (1/(D+1))*(Wd/Bd))^2
> 1-pt(MeanEst/sqrt(Td),df=nu)
```

```
[1] 0.05738632
```

This result suggests that there is marginal evidence for an effect of haplotype $h_8$ in `actn3` on the percentage change in the non-dominant arm muscle strength within Caucasians.                                                      □

## 5.2.2 A fully likelihood-based approach

In the previous section, we described two-stage approaches that involve first estimating posterior haplotype probabilities based on genotype information alone and second estimating the association between inferred haplotypes and a trait. Alternatively, we can apply a fully likelihood-based approach that involves simultaneous haplotype frequency estimation and estimation of haplotype–trait association. One marked advantage of this approach is that it uses information on the trait to inform us about the most likely haplotype pair for an individual. Consider a simple example in which an individual is ambiguous between the diplotypes $(h_1, h_4)$ and $(h_2, h_3)$. That is, based on this person's genotype, we are not sure whether $h_1$ and $h_4$ are present or $h_2$ and $h_3$ are present. In the previous section, we described methods for estimating the probabilities that the diplotype is $(h_1, h_4)$ or $(h_2, h_3)$ based on information about the population-level frequencies and underlying population genetic models. Now suppose that we have also measured a quantitative trait $y$ for this person and it is equal to 2.1. Suppose further that those individuals in our sample who are homozygous for $h_1$, $h_2$, $h_3$ or $h_4$ (and therefore fully observed) have an average response of $\bar{y}_{h_1} = 2.0$, $\bar{y}_{h_2} = -1.5$, $\bar{y}_{h_3} = -2.0$ and $\bar{y}_{h_4} = 2.5$, respectively. In this simple case and depending on the model of association, the ambiguous individual may be more likely to have the $(h_1, h_4)$ haplotype since the observed trait for this individual is closer to the respective sample averages.

Details of the fully likelihood-based approach are given in Schaid et al. (2003), Lake et al. (2003), Lin and Zeng (2006) and in the supplemental notes at the end of this chapter. Briefly, similar to the method described in Section 5.1.1, this approach involves the application of an expectation-maximization (EM) algorithm. In this setting, however, the complete data log likelihood is now defined in terms of both the haplotypes and the trait under investigation. The *E-step* involves taking the conditional expectation of this complete data log likelihood. This amounts to calculating a weighted sum of the likelihood evaluated for each possible haplotype pair, where the weights are the posterior probabilities of these pairs given the observed genotype and trait. This posterior probability is evaluated based on the current parameter estimates and is given formally by Equation (5.35) in the supplemental notes to this chapter. The *M-step* maximizes the conditional expectation derived in the E-step to arrive at new parameter estimates. Finally, the two steps are repeated until a convergence criterion is met. Testing of the resulting parameters requires incorporating the additional uncertainty resulting from the missingness in the data. We demonstrate this approach in the following example using the `haplo.glm()` function of the `haplo.stats` package. While methods for handling specific departures from HWE have been described, this function assumes HWE and should be applied within racial and ethnic strata as appropriate.

*Example 5.6 (EM for estimation and testing of haplotype–trait association).* In this example, we implement the fully likelihood-based approach to estimating haplotype frequencies and haplotype–trait association. Again we consider data from the FAMuSS study and focus on association between haplotypes within the `actn3` gene and the percentage change in the non-dominant arm muscle strength. Using the genotype data matrix `Geno.C` that we generated in Example 5.1, we can apply the following code, where again `haplo.glm()` is a function within the `haplo.stats` package:

```
> attach(fms)
> Geno.C <- setupGeno(Geno.C)
> Trait <- NDRM.CH[Race=="Caucasian" & !is.na(Race)]
> Dat <- data.frame(Geno.C=Geno.C, Trait=Trait)
> library(haplo.stats)
> haplo.glm(Trait~Geno.C,data=Dat,
+          allele.lev=attributes(Geno.C)$unique.alleles)

  Call:
haplo.glm(formula = Trait ~ Geno.C, data = Dat,
    allele.lev = attributes(Geno.C)$unique.alleles)

Coefficients:
             coef    se  t.stat    pval
(Intercept) 50.678 2.217 22.8572 0.0000
Geno.C.3     8.496 0.611 13.8975 0.0000
```

```
Geno.C.5      -0.441 7.280 -0.0606 0.9517
Geno.C.8       2.011 1.891  1.0633 0.2880
Geno.C.9       8.422 3.510  2.3995 0.0167
Geno.C.rare    3.985 6.294  0.6331 0.5268
```

```
Haplotypes:
             loc.1 loc.2 loc.3 loc.4 hap.freq
Geno.C.3       C     A     T     G    0.0125
Geno.C.5       C     G     C     G    0.0108
Geno.C.8       T     A     T     G    0.4024
Geno.C.9       T     G     C     A    0.0839
Geno.C.rare    *     *     *     *    0.0181
haplo.base     C     G     C     A    0.4722
```

By default, the base haplotype is set equal to the most prevalent one in our sample, in this case CGCA. Each of the $p$-values returned by applying the haplo.glm() function corresponds to a test that the effect of the corresponding haplotype, compared with this base haplotype, is equal to 0. For example, we see that the effect of Geno.C.9 is 8.422 with a corresponding $p$-value of 0.0167. This implies that the mean percentage change in muscle strength is 8.422 greater among individuals with one copy of the Geno.C.9 haplotype compared with individuals that are homozygous for the CGCA haplotype and this effect is significantly different from zero.

Several parameters can be controlled in the analysis above. First of all, we can change the base haplotype, which is useful if we are trying to compare results across different racial/ethnic strata within which the most prevalent haplotype differs. This is also useful if there is a specific haplotype to which we would like to compare all other haplotypes. To see how we do this, suppose we want TGCA to be the base haplotype. In this case, we use the following code, where the index $h_9$ is based on the output from Example 5.1:

```
> haplo.glm(Trait~Geno.C,data=Dat,
+                allele.lev=attributes(Geno.C)$unique.alleles,
+                control=haplo.glm.control(haplo.base=9))

  Call:
haplo.glm(formula = Trait ~ Geno.C, data = Dat,
    control = haplo.glm.control(haplo.base = 9),
    allele.lev = attributes(Geno.C)$unique.alleles)


Coefficients:
              coef   se  t.stat    pval
(Intercept) 67.5222 5.33 12.6714 0.0000
Geno.C.3     0.0738 4.60  0.0161 0.9872
Geno.C.4    -8.4221 3.14 -2.6808 0.0075
Geno.C.5    -8.8630 7.35 -1.2066 0.2279
Geno.C.8    -6.4110 3.08 -2.0842 0.0375
Geno.C.rare -4.4371 6.50 -0.6830 0.4948
```

```
Haplotypes:
          loc.1 loc.2 loc.3 loc.4 hap.freq
Geno.C.3       C     A     T     G   0.0125
Geno.C.4       C     G     C     A   0.4722
Geno.C.5       C     G     C     G   0.0108
Geno.C.8       T     A     T     G   0.4024
Geno.C.rare    *     *     *     *   0.0181
haplo.base     T     G     C     A   0.0839
```

Another important parameter that we can control is an indicator for the type of genetic model. By default, the `haplo.glm()` function assumes an additive genetic model, so that the effect of having two copies of a haplotype is twice the effect of having a single copy of the haplotype. Additive models as well as other potential genetic models are defined in Section 2.3.4. We can easily specify an alternative model structure by specifying the `haplo.effect` parameter within `haplo.glm.control`. For example, under a dominant genetic model in which one or more copies of a haplotype cause the effect, we have

```
> haplo.glm(Trait~Geno.C,data=Dat,
+          allele.lev=attributes(Geno.C)$unique.alleles,
+          control=haplo.glm.control(haplo.effect="dominant"))

  Call:
haplo.glm(formula = Trait ~ Geno.C, data = Dat,
    control = haplo.glm.control(haplo.effect = "dominant"),
    allele.lev = attributes(Geno.C)$unique.alleles)
```

```
Coefficients:
              coef   se t.stat     pval
(Intercept) 48.94 2.38 20.587 0.000000
Geno.C.3     4.35 1.25  3.478 0.000533
Geno.C.5     1.39 9.43  0.147 0.883253
Geno.C.8     4.58 2.70  1.699 0.089813
Geno.C.9    13.90 4.87  2.856 0.004412
Geno.C.rare  3.43 7.51  0.458 0.647324
```

```
Haplotypes:
          loc.1 loc.2 loc.3 loc.4 hap.freq
Geno.C.3       C     A     T     G   0.0125
Geno.C.5       C     G     C     G   0.0108
Geno.C.8       T     A     T     G   0.4024
Geno.C.9       T     G     C     A   0.0838
Geno.C.rare    *     *     *     *   0.0181
haplo.base     C     G     C     A   0.4724
```

We see that the results of this analysis are slightly different from the results under the additive assumption. Specifically, there is stronger evidence of an association between our trait and haplotypes Geno.C.3=CATG and Geno.C.9=TGCA.

Other parameters that can be controlled include handling of missing genotype data, the minimum prevalence for inclusion in the `rare` haplotype group and the distributional family. The default setting handles missing genotype information within the EM framework while excluding individuals with missing trait or covariate data. The `family` option allows testing association with a binary trait, similar to the `glm()` function. □

In general, the fully likelihood-based approach described in this section is preferable to the two-stage approaches described in Section 5.2.1 when the primary aim is to characterize haplotype–trait association. This approach is straightforward to implement and carries many desirable statistical properties. The two-stage approaches, however, have one primary advantage that is relevant to studies across multiple racial or ethnic groups. These approaches allow for estimation of haplotype effects within subgroups and then combining the data for estimation and testing of effects. For example, we can use the `haplo.em()` function within African Americans and Caucasians separately to arrive at posterior probability estimates of each individual and impute data based on these estimates. We are then able to fit a model using all of the data combined. Repeating the imputation procedure as described in Section 5.2.1 yields valid estimates and tests. By using all of the data across racial and ethnic groups in the estimation step, this approach will give us more power to detect associations under the assumption that there is no effect modification by race or ethnicity. The `haplo.glm()` function, on the other hand, makes the HWE assumption and is therefore most appropriately applied within racial or ethnic groups within which there is no evidence of a departure from HWE.

# Problems

**5.1.** Using the FAMuSS data, estimate the `resistin` haplotype frequencies for Caucasians and African Americans separately. Determine whether the frequency of the most common haplotype in Caucasians is statistically different from the frequency of this haplotype in African Americans.

**5.2.** Based on the HGDP data, estimate the `AKT1` haplotype frequencies within groups defined by the variable `Population`. Repeat estimation within groups defined by `Geographic.origin`. Compare and contrast your conclusions for the two analyses.

**5.3.** Is there an association between `Gender` and `Geographic.origin` in the HGDP data? If so, how would this influence your interpretation of an analysis of a genotype–trait association?

**5.4.** Apply haploytpe trend regression (HTR) to determine if there is an association between the `resistin` haplotypes and change in non-dominant arm muscle strength within African Americans using the FAMuSS data.

**5.5.** Using the expectation-maximization approach of the `haplo.glm()` function, determine if there is an association between the `resistin` haplotypes and change in non-dominant arm muscle strength, as measured by `NDRM.CH`, within African Americans, based on the FAMuSS data. Consider both dominant and additive genetic models.

## Supplemental notes

### EM approach to haplotype frequency estimation

Before stepping through this algorithm, we begin by defining our notation. Let $G_i$ represent the (observed) unphased genotype for individual $i$, where $i = 1, \ldots, n$. Further suppose $H_i$ is the (unobservable) combination of haplotypes for individual $i$. So, for example, in a diploid setting, $H_i$ represents a pair of haplotypes and multiple pairs are generally consistent with the observed genotype, $G_i$. We let $\mathcal{S}(G_i)$ denote the set of all $H_i$ that are consistent with $G_i$. The set of all $K$ single haplotypes observed within our sample of individuals is denoted $h_1, \ldots, h_K$, and the corresponding population-level frequencies are given by $\theta = (\theta_1, \ldots, \theta_K)$. The first goal of our analysis is to estimate these haplotype frequencies.

To begin, we define the complete data by $\mathbf{X}^c = [\mathbf{G}, \mathbf{H}]$, where $\mathbf{G} = (G_1, \ldots, G_n)$ and $\mathbf{H} = (H_1, \ldots, H_n)$. The complete data likelihood is then given by

$$L(\theta|\mathbf{X}^c) = \prod_{i=1}^{n} Pr(H_i|\theta) \tag{5.15}$$

where we assume the haplotype set probabilities, $Pr(H_i|\theta)$, are given by the multinomial distribution. That is,

$$Pr(H_i|\theta) = C_i \prod_{k=1}^{K} \theta_k^{\delta_{ik}} \tag{5.16}$$

where $C_i = 2/(\delta_{i1}! \ldots \delta_{iK}!)$, $\delta_{ik}$ is the number of copies of $h_k$ carried by individual $i$ and $\sum_{k=1}^{K} \theta_k = 1$. In a diploid population, $\delta_{ik}$ can take on the value 0, 1 or 2 and $C_i$ equals 1 if individual $i$ is homozygous and 2 otherwise. Taking the logarithm of the complete data likelihood given in Equation (5.15) yields

$$l_\theta = \log L(\theta|\mathbf{X}^c) = \sum_{i=1}^{n} \log Pr(H_i|\theta) \tag{5.17}$$

The *E-step* involves taking the conditional expectation of this complete data log likelihood based on the observed data and current estimates of the population haplotype frequencies. For the $t$th iteration of the EM algorithm, we denote this estimate by $\widehat{\theta}^{(t)}$. That is, we calculate

$$E\left(l_\theta|\mathbf{G}, \widehat{\theta}^{(t)}\right) = \sum_{i=1}^{n} \sum_{H_i \in \mathcal{S}(G_i)} \widehat{p}_{H_i}^{(t)} \log Pr(H_i|\theta) \tag{5.18}$$

where $\widehat{p}_{H_i}^{(t)}$ is the estimated posterior probability of $H_i$ given $G_i$ and $\widehat{\theta}^{(t)}$. Formally, we write

$$\widehat{p}_{H_i}^{(t)} = \frac{Pr(H_i|\widehat{\theta}^{(t)})}{\sum_{H_i \in \mathcal{S}(G_i)} Pr(H_i|\widehat{\theta}^{(t)})} \tag{5.19}$$

The *M-step* involves maximizing the conditional expectation of Equation (5.18) with respect to $\theta$. To do this, we set the first derivative equal to 0 and solve for $\theta$. Note that the partial derivative of this conditional expectation is given by

$$\frac{\partial E\left(l_\theta|\mathbf{G}, \widehat{\theta}^{(t)}\right)}{\partial \theta_k} = \sum_{i=1}^{n} \sum_{H_i \in \mathcal{S}(G_i)} \widehat{p}_{H_i}^{(t)} \partial \log Pr(H_i|\theta)/\partial \theta_k \tag{5.20}$$

To find $\partial \log Pr(H_i|\theta)/\partial \theta_k$, we first note that we can write

$$\log Pr(H_i|\theta) = \log C_i + \sum_{k=1}^{K} \delta_{ik} \log \theta_k \tag{5.21}$$

Taking the first derivative of this quantity yields

$$\frac{\partial \log Pr(H_i|\theta)}{\partial \theta_k} = \partial \left[\log C_i\right]/\partial \theta_k + \partial \left[\sum_{k=1}^{K} \delta_{ik} \log \theta_k\right]/\partial \theta_k \tag{5.22}$$

The first term in this sum is equal to 0 since the argument of the derivative does not involve $\theta$. Therefore, after introducing the constraint $\sum_{k=1}^{K} \theta_k = 1$, we have

$$\frac{\partial \log Pr(H_i|\theta)}{\partial \theta_k} = \sum_{k=1}^{K} \frac{\partial \left( \delta_{ik} \log \theta_k \right)}{\partial \theta_k}$$

$$= \sum_{k=1}^{K} \frac{\partial \left( \delta_{ik} \log \theta_k \right)}{\partial \theta_k} + \frac{\partial \left[ \delta_{iK} \log \left( 1 - \sum_{k=1}^{K-1} \theta_k \right) \right]}{\partial \theta_k} \quad (5.23)$$

$$= \frac{\delta_{ik}}{\theta_k} - \frac{\delta_{iK}}{\left( 1 - \sum_{k=1}^{K-1} \theta_k \right)} = \frac{\delta_{ik}}{\theta_k} - \frac{\delta_{iK}}{\theta_K}$$

We can therefore write Equation (5.20) as follows for each $k = 1, \ldots, K - 1$:

$$\frac{\partial E\left( l_\theta | \mathbf{G} \right)}{\partial \theta_k} = \sum_{i=1}^{n} \sum_{H_i \in \mathcal{S}(G_i)} \widehat{p}_{H_i}^{(t)} \left( \frac{\delta_{ik}}{\theta_k} - \frac{\delta_{iK}}{\theta_K} \right) \quad (5.24)$$

Setting these $K - 1$ equations equal to 0 yields

$$\frac{1}{\theta_k} \sum_{i=1}^{n} \sum_{H_i \in \mathcal{S}(G_i)} \widehat{p}_{H_i}^{(t)} \delta_{ik} = \frac{1}{\theta_K} \sum_{i=1}^{n} \sum_{H_i \in \mathcal{S}(G_i)} \widehat{p}_{H_i}^{(t)} \delta_{iK} \quad (5.25)$$

for each $k = 1, \ldots, K - 1$. Since the right-hand side of this equation is the same for all $k = 1, \ldots, K - 1$, we can write

$$\frac{1}{\widehat{\theta}_k} \sum_{i=1}^{n} \sum_{H_i \in \mathcal{S}(G_i)} \widehat{p}_{H_i}^{(t)} \delta_{ik} = \frac{1}{\widehat{\theta}_1} \sum_{i=1}^{n} \sum_{H_i \in \mathcal{S}(G_i)} \widehat{p}_{H_i}^{(t)} \delta_{i1} \quad (5.26)$$

for $k = 2, \ldots, K - 1$. Solving for $\widehat{\theta}_k$ yields

$$\widehat{\theta}_k = \frac{\sum_{i=1}^{n} \sum_{H_i \in \mathcal{S}(G_i)} \widehat{p}_{H_i}^{(t)} \delta_{ik}}{\sum_{i=1}^{n} \sum_{H_i \in \mathcal{S}(G_i)} \widehat{p}_{H_i}^{(t)} \delta_{i1}} \widehat{\theta}_1 \quad (5.27)$$

Thus we can derive an estimate of $\theta_1$ and use this to estimate $\theta_k$ for each $k = 2, \ldots, K - 1$.

An estimate of $\theta_1$ is derived based on Equation (5.25) for $k = 1$ and using the relationship $\theta_K = 1 - \sum_{k=1}^{K-1} \theta_k$. To see this, note that we can write

$$\widehat{\theta}_1 \sum_{i=1}^{n} \sum_{H_i \in \mathcal{S}(G_i)} \widehat{p}_{H_i}^{(t)} \delta_{iK} = \widehat{\theta}_K \sum_{i=1}^{n} \sum_{H_i \in \mathcal{S}(G_i)} \widehat{p}_{H_i}^{(t)} \delta_{i1}$$

$$= \left(1 - \sum_{k=1}^{K-1} \widehat{\theta}_k\right) \sum_{i=1}^{n} \sum_{H_i \in \mathcal{S}(G_i)} \widehat{p}_{H_i}^{(t)} \delta_{i1}$$

$$\text{(5.28)}$$

$$= \left(1 - \sum_{k=1}^{K-1} \frac{\sum_{i=1}^{n} \sum_{H_i \in \mathcal{S}(G_i)} \widehat{p}_{H_i}^{(t)} \delta_{ik}}{\sum_{i=1}^{n} \sum_{H_i \in \mathcal{S}(G_i)} \widehat{p}_{H_i}^{(t)} \delta_{i1}} \widehat{\theta}_1\right) \sum_{i=1}^{n} \sum_{H_i \in \mathcal{S}(G_i)} \widehat{p}_{H_i}^{(t)} \delta_{i1}$$

$$= \sum_{i=1}^{n} \sum_{H_i \in \mathcal{S}(G_i)} \widehat{p}_{H_i}^{(t)} \delta_{i1} - \widehat{\theta}_1 \sum_{k=1}^{K-1} \sum_{i=1}^{n} \sum_{H_i \in \mathcal{S}(G_i)} \widehat{p}_{H_i}^{(t)} \delta_{ik}$$

The next to final step in the equation above involves simply plugging in the result of Equation (5.27) for $k = 2, \ldots, K - 1$. This is equivalent to

$$\widehat{\theta}_1 \sum_{i=1}^{n} \sum_{H_i \in \mathcal{S}(G_i)} \widehat{p}_{H_i}^{(t)} \sum_{k=1}^{K} \delta_{ik} = \sum_{i=1}^{n} \sum_{H_i \in \mathcal{S}(G_i)} \widehat{p}_{H_i}^{(t)} \delta_{i1} \qquad \text{(5.29)}$$

Solving for $\widehat{\theta}$ and using the identities $\sum_{k=1}^{K} \delta_{ik} = 2$ and $\sum_{H_i \in \mathcal{S}(G_i)} \widehat{p}_{H_i}^{(t)} = 1$, we have

$$\widehat{\theta}_1^{(t+1)} = \frac{1}{2n} \sum_{i=1}^{n} \sum_{H_i \in \mathcal{S}(G_i)} \widehat{p}_{H_i}^{(t)} \delta_{i1} \qquad \text{(5.30)}$$

Finally, plugging this result back into Equation (5.27) gives the general result

$$\widehat{\theta}_k^{(t+1)} = \frac{1}{2n} \sum_{i=1}^{n} \sum_{H_i \in \mathcal{S}(G_i)} \widehat{p}_{H_i}^{(t)} \delta_{ik} \qquad \text{(5.31)}$$

In summary, the EM algorithm involves first calculating individual-level posterior probabilities for each of the haplotype pairs that are consistent with the observed genotypes. This is given by Equation (5.19). The second step then involves updating the parameter values based on the current posterior probabilities. These new parameter estimates are given in Equation (5.31). We then iterate between these two steps until the value of the estimate does not vary substantially from one iteration to the next. For example, we may stipulate that iteration continues until $\max_k \left(|\widehat{\theta}_k^{(t)} - \widehat{\theta}_k^{(t+1)}|/\widehat{\theta}_k^{(t)}\right) < 1.0 \times 10^{-5}$. Alternatively, we can base our convergence criterion on the observed data likelihood, given by $\prod_{i=1}^{n} Pr(G_i|\theta) = \prod_{i=1}^{n} \sum_{H_i \in S(G_i)} Pr(H_i|\theta)$.

## EM approach to haplotype–trait association

Consider the general linear model setting in which the dependent variable, $Y$, arises from an exponential family with a canonical link function. Let the covariates and haplotype information be denoted collectively by $[\mathbf{X}\ \mathbf{H}]$. Then we have the density of $y$ given by

$$f_\beta(y|\mathbf{X}, \mathbf{H}) = \exp\left\{\frac{y[\mathbf{X}^T\ \mathbf{H}^T]\beta - b([\mathbf{X}^T\ \mathbf{H}^T]\beta)}{a(\Psi)} + c(y, \Psi)\right\} \quad (5.32)$$

where $a$, $b$ and $c$ are known functions, $\beta$ is a vector of parameters corresponding to the design matrix $X$ and $\Psi$ represents a scale parameter, such as the error variance.

In general, haplotype information is not observed and estimation of $\beta$ can again proceed using the EM algorithm. Recall that this is an iterative two-stage approach involving first taking the conditional expectation of the complete data log likelihood (E-step) and then maximizing this with respect to the parameters (M-step). This process is then repeated until a convergence criterion is met. The complete data are now given by $Y$ and $H$, and the complete data likelihood for the $i$ individual can be written

$$L_i^c(\Phi) = L_i^c(\Phi|Y_i, \mathbf{X}_i, \mathbf{H}_i) = f(Y_i, \mathbf{H}_i|\mathbf{X}_i, \Phi) = f_\beta(Y_i|\mathbf{X}_i, \mathbf{H}_i)Pr_\theta(\mathbf{H}_i) \quad (5.33)$$

where $\Phi = (\beta, \theta)$, $\theta = (\theta_1, \ldots, \theta_K)$ is a vector with $k$th element equal to the population-level frequency of haplotype $k$.

The conditional expectation is given by the weighted sum of this likelihood given each possible haplotype, where the weights are equal to the posterior probabilities of the haplotypes given the observed genotypes. Formally, we write

$$E[\log L_i^c(\Phi)|Y_i, \mathbf{X}_i, \mathbf{G}_i] = \sum_{H \in \mathcal{S}(G_i)} p_i(\Phi)\left[\log f_\beta(Y_i|\mathbf{X}_i, \mathbf{H}_i) + \log Pr_\theta(\mathbf{H}_i)\right]$$

$$(5.34)$$

where

$$p_i(\Phi) = \frac{f_\beta(Y_i|\mathbf{X}_i, \mathbf{H}_i)Pr_\theta(\mathbf{H}_i)}{\sum_{H \in \mathcal{S}(G_i)} f_\beta(Y_i|\mathbf{X}_i, \mathbf{H}_i)Pr_\theta(\mathbf{H}_i)} \quad (5.35)$$

For each iteration of the EM algorithm, we substitute the current estimates of $\beta$ and $\theta$ into the calculation of the posterior haplotype probabilities (Equation (5.35)) and then update these estimates by maximizing the conditional expectation (Equation (5.34)). Maximization at the $(t+1)$st iteration proceeds by taking the partial derivative with respect to each parameter, setting this equal to 0

$$\sum_{i=1}^{n} \sum_{H \in \mathcal{S}(G_i)} \widehat{p}_i^{(t)} \frac{\partial \log f_\beta(Y_i | \mathbf{X}_i, \mathbf{H}_i)}{\partial \beta} = 0 \tag{5.36}$$

$$\sum_{i=1}^{n} \sum_{H \in \mathcal{S}(G_i)} \widehat{p}_i^{(t)} \frac{\partial \log Pr_\theta(\mathbf{H}_i)}{\partial \theta} = 0 \tag{5.37}$$

and solving for $\beta$ and $\theta$, respectively. Here note that the posterior probabilities are evaluated at the current parameter values, so that $\widehat{p}_i^{(t)} = p_i(\widehat{\Phi}^{(t)})$.

Under the HWE assumption, we can write the probability of $H = (h_k, h_l)$ as the product of the probability of each element of the pair, as given in Equation (5.16). Notably, the HWE assumption is not necessary, and Lin and Zeng (2006) describe estimation under specific departures from HWE. Drawing inference about $\beta$ and $\theta$ in this setting requires estimation of the variance/covariance matrix. This matrix is given by the inverse of the observed information matrix, which can be approximated with the empirical observed information (Meilijson, 1989). Formally, we write

$$I(\widehat{\Phi}) = \sum_{i=1}^{n} s_i(\widehat{\Phi}) s_i(\widehat{\Phi})^T \tag{5.38}$$

where $s_i(\widehat{\Phi})$ is given by the inside sums of Equations (5.36) and (5.37), evaluated at the last iteration of the EM algorithm (McLachlan and Krishnan, 1997).

## Supplemental R scripts

```
######################################################################
# Description: This function creates a design matrix with i,j
#              element equal to the conditional expectation
#              of the number of copies of haplotype j for
#              individual i based on the output from haplo.em()
# Input:       HaploEM (object resulting from haplo.em())
# Output:      XmatHap
######################################################################


HapDesign <- function(HaploEM){
    Nobs <- length(unique(HaploEM$indx.subj)) #number of observations
    Nhap <- length(HaploEM$hap.prob)          #number of haplotypes
    XmatHap <- matrix(data=0,nrow=Nobs,ncol=Nhap)
    for (i in 1:Nobs){
        IDSeq <- seq(1:sum(HaploEM$nreps))[HaploEM$indx.subj==i]
        for (j in 1:length(IDSeq)){
                XmatHap[i,HaploEM$hap1code[IDSeq][j]] <-
```

```
                    XmatHap[i,HaploEM$hap1code[IDSeq][j]] +
                    HaploEM$post[IDSeq][j]
                XmatHap[i,HaploEM$hap2code[IDSeq][j]] <-
                    XmatHap[i,HaploEM$hap2code[IDSeq][j]] +
                    HaploEM$post[IDSeq][j]
            }
        }
    return(XmatHap)
    }


##########################################################################
# Description: This function creates a vector with jth element
#              equal to the standard error of haplotype j
#              based on the output from haplo.em()
# Input:       HaploEM (object resulting from haplo.em())
# Output:      HapSE
##########################################################################


HapFreqSE <- function(HaploEM){
    HapMat <- HapDesign(HaploEM)
    Nobs <- length(unique(HaploEM$indx.subj)) #number of observations
    Nhap <- length(HaploEM$hap.prob)          #number of haplotypes
    S.Full<-matrix(data=0, nrow=Nobs, ncol=Nhap-1)
        for(i in 1:Nobs){
                for(k in 1:(Nhap-1)){
                S.Full[i,k]<-HapMat[i,k]/HaploEM$hap.prob[k]-
                    HapMat[i,Nhap]/HaploEM$hap.prob[Nhap]
                }
    }
    Score<-t(S.Full)%*%S.Full
    invScore<-solve(Score)
    HapSE<-c(sqrt(diag(invScore)),
            sqrt(t(rep(1,Nhap-1))%*%invScore%*%rep(1,Nhap-1)))
    return(HapSE)
    }
```

# 6

# Classification and Regression Trees

Classification and regression trees (CARTs) are an approach to discovering relationships among a large number of independent (predictor) variables and a categorical or continuous trait. Classification trees are applied to categorical outcomes, while regression trees apply to continuous traits. Both involve the application of a recursive algorithm that aims to partition individuals into groups in a way that minimizes the within-group heterogeneity. CART was originally described by Breiman *et al.* (1993) and has gained popularity in recent years as a method for identifying structure in high-dimensional data settings. In the following sections, we begin by describing methods for constructing a tree. This involves defining a measure of heterogeneity, or what is commonly referred to as *node impurity*, as well as determining how predictor variables are input into the model. Both of these components will impact the resulting tree and need to be considered and defined carefully to reflect the scientific questions at hand. We then describe methods for refining this tree to arrive at a final reproducible model. Further discussions of CART methods can be found in Breiman *et al.* (1993) and Zhang and Singer (1999). In Chapter 7, we describe extensions of the CART model, including random forests and logic regression trees that offer some additional advantages.

## 6.1 Building a tree

### 6.1.1 Recursive partitioning

Suppose our data consist of a trait $\mathbf{y} = (y_1, \ldots, y_n)$ and $p$ potential predictor variables given by $\mathbf{x}_1, \ldots, \mathbf{x}_p$, where $\mathbf{x}_j = (x_{1j}, \ldots, x_{nj})^T$ and $i = 1, \ldots, n$ indicates individual. In general, interest lies in discovering the relationship between $\mathbf{X} = [\mathbf{x}_1, \ldots, \mathbf{x}_p]$ and $\mathbf{y}$. A tree is constructed by first determining the variable $\mathbf{x}_j$, among the set of all potential predictors, that is *most predictive* of the trait $\mathbf{y}$. The phrase "most predictive" is used loosely here and is defined explicitly in Section 6.1.2 below. Individuals in our sample are then

A.S. Foulkes, *Applied Statistical Genetics with R: For Population-based Association Studies*, Use R, DOI: 10.1007/978-0-387-89554-3_6,
© Springer Science+Business Media LLC 2009

divided into two groups based on their corresponding value of $\mathbf{x}_j$. Suppose, for example, that $\mathbf{x}_j$ is a binary indicator for the presence of the variant allele at the $j$th SNP under investigation. If $\mathbf{x}_j$ is identified as the most predictive variable, then individuals with at least one variant allele at this SNP are assigned to one group, while individuals that are homozygous wildtype are assigned to a second group. This process is then repeated within each of the two resulting groups.

More formally, let the set of all individuals be denoted $\Omega$. For simplicity of presentation, suppose all potential predictors are binary, and let the most predictive variable be given by $\mathbf{x}_{(1)}$. Individuals are first divided into $\Omega_1$ and $\Omega_2$ based on the value of $\mathbf{x}_{(1)}$. That is, we define $\Omega_1 = \{i : x_{i(1)} = 0\}$ and $\Omega_2 = \{i : x_{i(1)} = 1\}$ for $i = 1, \ldots, n$, representing individual. The next step of the tree-building algorithm involves again identifying the variable that is most predictive of $y_i$ but now within each of the groups of individuals given by $\Omega_1$ and $\Omega_2$. Suppose this variable is $\mathbf{x}_{(2)}$ for $\Omega_1$ and $\mathbf{x}_{(3)}$ for $\Omega_2$. Further subgroups are then defined based on the values of $\mathbf{x}_{(2)}$ and $\mathbf{x}_{i(3)}$. That is, we define

$$
\begin{aligned}
\Omega_{1,1} &= \{i : i \in \Omega_1 \text{ and } x_{i(2)} = 0\} = \{i : x_{i(1)} = 0 \text{ and } x_{i(2)} = 0\} \\
\Omega_{1,2} &= \{i : i \in \Omega_1 \text{ and } x_{i(2)} = 1\} = \{i : x_{i(1)} = 0 \text{ and } x_{i(2)} = 1\} \\
\Omega_{2,1} &= \{i : i \in \Omega_2 \text{ and } x_{i(3)} = 0\} = \{i : x_{i(1)} = 1 \text{ and } x_{i(3)} = 0\} \\
\Omega_{2,2} &= \{i : i \in \Omega_2 \text{ and } x_{i(3)} = 1\} = \{i : x_{i(1)} = 1 \text{ and } x_{i(3)} = 1\}
\end{aligned}
\tag{6.1}
$$

This partitioning procedure is then repeated recursively until a stopping rule is achieved. For example, we might continue splitting into smaller and smaller subgroups until a subgroup has less than five individuals. Note that the stopping rule may result in an asymmetric tree.

A visual representation of this recursive algorithm is given in Figure 6.1. Each circle is called a *node*. The first node is termed the *root node* or *parent node*, and subsequent splits result in the formation of left and right *daughter nodes*. A node is simply a representation of a group of individuals, which we denote with $\Omega$. Importantly, at each stage of the splitting algorithm, we are identifying the variable that is *most* predictive of the trait of interest among all variables under consideration; however, in general, this is not based on a formal statistical test. That is, these variables may not be significantly associated with the trait, even though they are more predictive than any other variable. Therefore, after building a tree in this manner, it is important to "prune" it back in order to avoid overfitting. In the following section, we describe criteria that are used for splitting nodes and then discuss pruning techniques and appropriate measures of tree accuracy in Section 6.2.

### 6.1.2 Splitting rules

In order to arrive at a tree structure similar to the one illustrated in Figure 6.1, first we need to define a rule for splitting our observations into subgroups. This

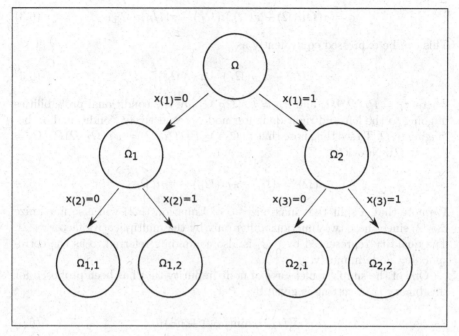

**Fig. 6.1.** Tree structure

typically proceeds by defining a measure of *node impurity*, or heterogeneity, and choosing the splitting variable to be the one that maximizes the reduction in this measure. While this is not the only approach for generating a tree, it is widely accepted and will be the focus of the presentation in this text. Let the impurity for node $\Omega$ be given by $I(\Omega)$. Formally, the goal is to choose the split that maximizes

$$\phi = I(\Omega) - I(\Omega_L) - I(\Omega_R) \tag{6.2}$$

where $\Omega_L$ and $\Omega_R$ are the left and right daughter nodes of $\Omega$, respectively. Several measures of impurity, $I(\Omega)$, are reasonable, and Breiman *et al.* (1993) suggest that in many instances the final tree tends to be insensitive to the choice of measure. In this section, we describe a few measures that are commonly applied in the context of binary and quantitative traits.

*Binary trait*

First consider a binary trait, $y$, that takes on the values 0 and 1. For example, $y$ may be an indicator for the presence of disease. In this case, we define $I(\Omega) = \pi(\Omega)i(\Omega)$ where $\pi(\Omega)$ is the probability of belonging to $\Omega$, and rewrite Equation (6.2) as

$$\phi = \pi(\Omega)i(\Omega) - \pi(\Omega_L)i(\Omega_L) - \pi(\Omega_R)i(\Omega_R) \tag{6.3}$$

This can be expressed equivalently as

$$\phi = [i(\Omega) - \pi_L i(\Omega_L) - \pi_R i(\Omega_R)]\,\pi(\Omega) \tag{6.4}$$

where $\pi_L = Pr(\Omega_L|\Omega)$ and $\pi_R = Pr(\Omega_R|\Omega)$ are the conditional probabilities of going to the left and right daughter nodes, respectively, conditional on belonging to $\Omega$. To see this, note that $\pi(\Omega_L) = Pr(\Omega_L, \Omega) = Pr(\Omega_L|\Omega)Pr(\Omega) = \pi_L \times \pi(\Omega)$. Now let

$$\triangle i(\Omega) = i(\Omega) - \pi_L i(\Omega_L) - \pi_R i(\Omega_R) \tag{6.5}$$

We note that a split that maximizes $\phi$ of Equation (6.2) will also maximize $\triangle i(\Omega)$ since these two functions differ only by the multiplicative factor $\pi(\Omega)$. The quantity represented by $i(\Omega)$ is also commonly referred to as the corresponding node impurity.

One of the simplest measures of node impurity that has been proposed for this binary trait setting is given by

$$i(\Omega) = \min(p_\Omega, 1 - p_\Omega) \tag{6.6}$$

where $p_\Omega = Pr(y = 1|\Omega)$ is the probability of being a case given membership in node $\Omega$. This impurity measure is commonly referred to as the *Bayes error*, *minimum error* or *misclassification cost*. Consider for example a setting in which there are equal numbers of cases and controls so that at the root node we have $i(\Omega) = \min(0.50, 1 - 0.50) = 0.50$. The idea behind this impurity measure is that a non-informative split would result in a node within which exactly half of the members are cases and half are controls. In this case, we would say the node is highly impure. On the other hand, if 80% of the individuals are cases and 20% controls (that is, $p_\Omega = 0.80$), then the node is relatively homogeneous. In other words, the impurity $i(\Omega) = 0.20$ is relatively low. While intuitively appealing, the minimum error has some practical limitations. These drawbacks include that it is common for there to be no single best split from a given node. In addition, the criterion does not favor trees with further growth potential. A complete discussion of these limitations is provided in Breiman *et al.* (1993).

Another commonly used measure of impurity is the *Gini index*, also called the *nearest neighbor error*, which is defined as

$$i(\Omega) = 2p_\Omega(1 - p_\Omega) \tag{6.7}$$

where $p_\Omega$ is again the probability of being a case, conditional on belonging to $\Omega$. Intuitively, the Gini index is also appealing since it is simply the sum of variances of Bernoulli random variables. To see this, note that for $y$ distributed Bernoulli($p$), so that $y$ takes on the value 1 with probability $p$ and 0 with

**Table 6.1.** Sample case–control data by genotype indicators

|          | $x_1 = 1$ | $x_1 = 0$ | $x_2 = 1$ | $x_2 = 0$ | $x_3 = 1$ | $x_3 = 0$ | Total |
|----------|-----------|-----------|-----------|-----------|-----------|-----------|-------|
| Cases:   | 60        | 40        | 80        | 20        | 50        | 50        | 100   |
| Controls:| 40        | 60        | 20        | 80        | 50        | 50        | 100   |
| Total:   | 100       | 100       | 100       | 100       | 100       | 100       | 200   |

probability $1 - p$, the variance of $y$ is given by $\text{Var}[y] = p(1 - p)$. Since $y$ can take on two values (0 or 1), we multiply this quantity by 2 to get the impurity measure given in Equation (6.7). More generally, for a categorical trait $y$ that takes on the values $1, \ldots, m$, the Gini index is given by

$$i(\Omega) = \sum_{r \neq s} p(r|\Omega)p(s|\Omega) \qquad \cdot (6.8)$$

where $p(r|\Omega)$ is the probability of being in class $r$ among individuals who fall into the node denoted $\Omega$, where $r = 1, \ldots, m$.

*Example 6.1 (Applying the Gini index).* Consider for example a prospective study in which there are exactly 100 cases and 100 controls. Further suppose there are three potential predictor variables $x_1$, $x_2$ and $x_3$ and a binary trait $y$. For simplicity of presentation, let $x_j$ be a vector of indicator variables for the presence of a variant allele at the $j$th SNP under investigation for all individuals in our sample. The overall impurity based on the Gini index at the root node is given by $i(\Omega) = 2p(1 - p) = 2(0.5^2) = 0.50$. The recursive partitioning algorithm begins by considering all splits of our data into two groups based on the value of $x_j$ for $j = 1, 2, 3$. Let us suppose our data are as given in Table 6.1.

From these tables, we see that $x_2$ does a better job at distinguishing between cases and controls than the other variables shown. This is described formally by recording a measure of the reduction in the within-node impurity corresponding to each variable. Let $\Omega_L(x_j)$ and $\Omega_R(x_j)$ be the left and right daughter nodes resulting from splitting based on the value of $x_j$ and let $\phi_j$ be the corresponding reduction in node impurity resulting from this split. We have

$$\begin{aligned} \phi_1 &= i(\Omega) - 0.5\left[i(\Omega_L(x_1))\right] - 0.5\left[i(\Omega_R(x_1))\right] \\ &= 0.50 - (0.6)(0.4) - (0.4)(0.6) = 0.02 \end{aligned} \qquad (6.9)$$

while $\phi_2 = 0.18$ and $\phi_3 = 0$. As expected, the variable that maximizes the reduction in impurity as measured by $\phi$ is $x_2$. Based on this information, we would begin by splitting our data into two groups based on whether a variant allele is present at SNP 2. This process of selecting the best variable is then repeated for each of the resulting daughter nodes. □

While generally acceptable in practice, the Gini index does have some drawbacks as well. Specifically, it tends to favor splits that result in two nodes with highly unbalanced sample sizes. Another popular alternative is the *entropy* function, also referred to as the *information index* and *deviance*, given by

$$i(\Omega) = -p_\Omega \log(p_\Omega) - (1 - p_\Omega) \log(1 - p_\Omega) \qquad (6.10)$$

where again we define $p_\Omega$ as the conditional probability of being a case, given membership in node $\Omega$. A motivation for this measure arises from likelihood theory. Suppose again that $y \sim \text{Bernoulli}(p)$ so that the log likelihood for our data is given by

$$\log L(p|n_1, n_2) = n_1 \log p + n_2 \log(1 - p) \qquad (6.11)$$

where $n_1$ and $n_2$ are the number of cases and controls, respectively. This function is maximized at

$$n_1 \log\left(\frac{n_1}{n}\right) + n_2 \log\left(\frac{n_2}{n}\right) = -ni(\Omega) \qquad (6.12)$$

where $i(\Omega)$ is the entropy function as defined in Equation (6.10). In practice, the entropy and Gini indexes tend to result in very similar trees for the case of a binary trait.

In the following example, we demonstrate how to construct a classification tree. Both the `rpart()` function of the `rpart` package and the `tree()` function of the `tree` package allow us to specify the measure of node impurity for tree fitting. The Gini index is the default, while the information index is given as an option.

*Example 6.2 (Creating a classification tree).* In this example, we return to the Virco data described in Section 1.3.3. Suppose we are interested in identifying polymorphisms within the protease region of the viral genome that are associated with greater *in vitro* sensitivity to nelfinavir (NFV) than indinavir (IDV). That is, we aim to characterize the relationship between mutations in the variables labeled P1, ..., P99 and a binary indicator for whether IDV.Fold is greater than NFV.fold. In this case, we begin by defining a dataframe, entitled `VircoGeno`, that contains binary indicators for the presence of a mutation at each of the 99 sites within the protease region:

```
> attach(virco)
> VircoGeno <- data.frame(virco[,substr(names(virco),1,1)=="P"]!="-")
```

We then define our trait and construct a classification tree as follows. Note that we define our trait as a `factor` variable since we are interested in creating a classification tree. We additionally specify `method="class"`, though this is the default setting when the trait is a factor:

```
> Trait <- as.factor(IDV.Fold > NFV.Fold)
> library(rpart)
> ClassTree <- rpart(Trait~., method="class", data=VircoGeno)
> ClassTree

n=976 (90 observations deleted due to missingness)

node), split, n, loss, yval, (yprob)
      * denotes terminal node

  1) root 976 399 FALSE (0.5911885 0.4088115)
    2) P54< 0.5 480 130 FALSE (0.7291667 0.2708333)
      4) P76< 0.5 466 116 FALSE (0.7510730 0.2489270) *
      5) P76>=0.5 14   0 TRUE (0.0000000 1.0000000) *
    3) P54>=0.5 496 227 TRUE (0.4576613 0.5423387)
      6) P46< 0.5 158  57 FALSE (0.6392405 0.3607595)
       12) P1< 0.5 115  31 FALSE (0.7304348 0.2695652) *
       13) P1>=0.5 43  17 TRUE (0.3953488 0.6046512) *
      7) P46>=0.5 338 126 TRUE (0.3727811 0.6272189)
       14) P10< 0.5 22   7 FALSE (0.6818182 0.3181818) *
       15) P10>=0.5 316 111 TRUE (0.3512658 0.6487342)
         30) P48< 0.5 278 106 TRUE (0.3812950 0.6187050)
           60) P20< 0.5 113  55 TRUE (0.4867257 0.5132743)
            120) P76< 0.5 92  40 FALSE (0.5652174 0.4347826) *
            121) P76>=0.5 21   3 TRUE (0.1428571 0.8571429) *
           61) P20>=0.5 165  51 TRUE (0.3090909 0.6909091) *
         31) P48>=0.5 38   5 TRUE (0.1315789 0.8684211) *
```

A visual representation of this output is generated with the `plot()` and `text()` functions and illustrated in Figure 6.2:

```
> plot(ClassTree)
> text(ClassTree)
```

Based on this output, first we see that the overall number of observations contributing to this analysis is given by n=976. The numbers and characters in each row of the output are interpreted as follows. The first number represents a node, and the subsequent expression, entitled split, defines this node. For example, consider the row denoted 2, in which we have P54 < 0.5. This indicates that viral particles that are wildtype at site P54 (i.e., P54=0) are assigned to the second node. The indentations in the output indicate daughter nodes, so that node 4 consists of sequences that match the criteria for both nodes 2 and 4. In other words, sequences in this node are wildtype at both P54 and P76. The next number, denoted n, is the number of viral particles within the corresponding node. For example, we have n=480 sequences that are wildtype at P54 and n=496 sequences that are mutant at this site, as indicated by the rows labeled 2 and 3. The next number, denoted loss, is the number of observations for which the trait is predicted incorrectly if we let this predicted value be the majority value within the node. This predicted

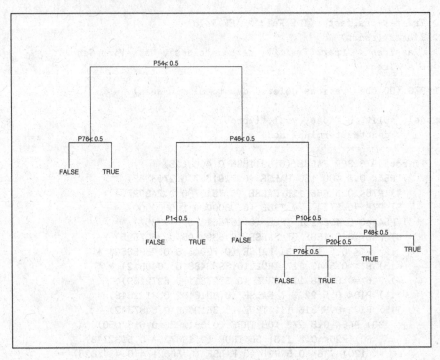

**Fig. 6.2.** Classification tree for Example 6.2

value is given by the following character string. For example, in the second node, the predicted value is FALSE, or equivalently IDV.Fold < NFV.Fold, and we see that loss=130 viral particles are predicted incorrectly. Finally, the corresponding estimated probabilities for each value of the trait are given in parentheses. For example, in the second node, we see that the proportion of sequences for which IDV.Fold > NFV.Fold is 130/480 = 0.27, while the proportion for which this does not hold is $1 - 0.27 = 0.73$.

By looking at the rows denoted 2) and 3), we see that the most predictive variable of greater NFV sensitivity is P54. Among viruses with a mutation at this site, indicated by P54 > 0.5, the estimated probability that IDV.Fold is greater than NFV.Fold is 0.54. On the other hand, this probability is 0.27 for viruses that are wildtype at this site. The next split in the tree is described in rows 4) and 5) of the output. Here we see that, among viruses that are wildtype at site P54, the most predictive variable is P76. In this case, the estimated probability of greater IDV than NFV fold resistance is 0.25 among viruses that are additionally wildtype at P76. On the other hand, this probability estimate is 1 for viruses that are mutant at P76. The final nodes, also referred to as *terminal* nodes, are indicated with an asterisk in the output. Importantly, these are the most predictive variables, but we have not yet assessed whether they are statistically significant or simply chance findings.

Several additional parameters can be specified within the `rpart()` function. A comprehensive description of these options is given in the associated documentation. In this and subsequent examples, we provide a brief discussion of a few that may be particularly useful. First, we can specify the split criterion to be used. The default criterion is the Gini index, and the information criterion is given as an alternative option. To specify this, we use the following code:

```
> rpart(Trait~., method="class", parms=list(split='information'),
+       data=VircoGeno)

n=976 (90 observations deleted due to missingness)

node), split, n, loss, yval, (yprob)
    * denotes terminal node

 1) root 976 399 FALSE (0.5911885 0.4088115)
   2) P54< 0.5 480 130 FALSE (0.7291667 0.2708333)
     4) P76< 0.5 466 116 FALSE (0.7510730 0.2489270) *
     5) P76>=0.5 14   0 TRUE (0.0000000 1.0000000) *
   3) P54>=0.5 496 227 TRUE (0.4576613 0.5423387)
     6) P46< 0.5 158  57 FALSE (0.6392405 0.3607595)
      12) P1< 0.5 115  31 FALSE (0.7304348 0.2695652) *
      13) P1>=0.5 43  17 TRUE (0.3953488 0.6046512) *
     .7) P46>=0.5 338 126 TRUE (0.3727811 0.6272189)
      14) P48< 0.5 299 120 TRUE (0.4013378 0.5986622)
        28) P20< 0.5 125  62 FALSE (0.5040000 0.4960000)
          56) P76< 0.5 104  44 FALSE (0.5769231 0.4230769) *
          57) P76>=0.5 21   3 TRUE (0.1428571 0.8571429) *
        29) P20>=0.5 174  57 TRUE (0.3275862 0.6724138) *
      15) P48>=0.5 39   6 TRUE (0.1538462 0.8461538) *
```

Notably, this results in a different tree than we saw above, though the first several splits are identical. Since splits that are lower down in the tree tend to be pruned back, as we describe in Section 6.2 below, these differences may not be relevant.

We are also able to specify relevant criteria regarding the numbers of observations within nodes. Specifically, the control parameter `minsplit` allows us to specify the number of observations that are required in a node in order to consider additional splits. The parameter labeled `minbucket` additionally allows us to indicate the minimum number of observations required in a terminal node. The two parameters are specified as follows:

```
> rpart(Trait~., method="class", parms=list(split='gini'),
+       control=rpart.control(minsplit=150, minbucket=50),
+       data=VircoGeno)

n=976 (90 observations deleted due to missingness)
```

```
node), split, n, loss, yval, (yprob)
     * denotes terminal node

1) root 976 399 FALSE (0.5911885 0.4088115)
  2) P54< 0.5 480 130 FALSE (0.7291667 0.2708333) *
  3) P54>=0.5 496 227 TRUE (0.4576613 0.5423387)
    6) P46< 0.5 158  57 FALSE (0.6392405 0.3607595) *
    7) P46>=0.5 338 126 TRUE (0.3727811 0.6272189) *
```

Comparing this with the original output in which we applied the Gini index, we see that further splits of nodes 2, 6 and 7 are no longer included. This is due to the fact that splits at these nodes all result in daughter nodes that contain less than the required number of observations, given by `minbucket=50`. We return to consideration of additional options for the `rpart()` function in subsequent examples. □

*Quantitative trait*

Now suppose the trait or outcome of interest is a quantitative measure of disease progression. In this case, the most commonly used measure of node impurity is the average sum of squared deviations from the mean, also referred to as the mean square error (MSE) and given by

$$I(\Omega) = \frac{1}{n_\Omega} \sum_{i \in \Omega} (y_i - \bar{y})^2 \qquad (6.13)$$

where $\bar{y}$ is the mean trait across our sample. Notably, this is the same as the least squares criterion used to find coefficient estimates in a linear regression, as described in Section 2.2.3. For a given binary split, the impurities for the left and right daughter nodes are given respectively by $I(\Omega_L) = 1/n_L \sum_{i \in \Omega_L} (y_i - \bar{y}_L)^2$ and $I(\Omega_R) = 1/n_R \sum_{i \in \Omega_R} (y_i - \bar{y}_R)^2$, where $\bar{y}_L$ and $\bar{y}_R$ are the corresponding mean responses and $n_L$ and $n_R$ are the sample sizes in $\Omega_L$ and $\Omega_R$, respectively. An example of fitting a regression tree is given below.

*Example 6.3 (Generating a regression tree).* Consider the Virco data of Example 6.2 and again suppose we are interested in identifying mutations in the protease region that are associated with a difference in IDV and NFV fold resistance. In this case, we define a quantitative trait as the difference in these two values:

```
> attach(virco)
> Trait <- NFV.Fold - IDV.Fold
```

Using the `VircoGeno` matrix created in Example 6.2, we fit a regression tree as follows:

```
> library(rpart)
> RegTree <- rpart(Trait~., method="anova", data=VircoGeno)
> RegTree
```

```
n=976 (90 observations deleted due to missingness)

node), split, n, deviance, yval
     * denotes terminal node

 1) root 976 6437933.00    4.288320
   2) P54>=0.5 496 1247111.00  -3.916935
     4) P46>=0.5 338   343395.20 -10.567160 *
     5) P46< 0.5 158   856789.90  10.309490
      10) P58< 0.5 144   110944.10    2.570139 *
      11) P58>=0.5 14   648503.60   89.914290 *
   3) P54< 0.5 480 5122921.00   12.767080
     6) P73< 0.5 422   145579.90    5.706635 *
     7) P73>=0.5 58 4803244.00   64.137930
      14) P35< 0.5 45    26136.17    8.171111 *
      15) P35>=0.5 13 4148242.00  257.869200 *
```

Notably here we specify `method="anova"`, although this is the default for a numeric trait. Based on this output, we again see that `n=976` observations contributed to this analysis. In this setting, `yval` corresponds to the mean trait or predicted value for observations within the corresponding node. For example, at the `root` node, the mean difference in NFV and IDV fold resistance is 4.29. Among sequences that are mutant at `P54`, this mean difference is $-3.92$, while sequences that are wildtype at `P54` are predicted to have a difference of 12.77. The `deviance` is defined as the sum of the squared differences between the observed trait and the predictive value over all observations within the corresponding node. Dividing this quantity by the within-node value of n yields $I(\Omega)$ of Equation (6.13).  □

### 6.1.3 Defining inputs

In Section 6.1.2 above, we focus on settings in which the predictor variables are binary. More generally, in the genetic association setting, we have a set of potential genetic predictor variables that are categorical as well as several clinical and demographic covariates that are categorical and continuous. The covariates could be, for example, gender, smoking status, race, weight, height, etc. In this section, we begin by describing how multilevel categorical and continuous variables are handled as inputs in a binary splitting tree. We then give specific attention to covariates and how to incorporate them into tree fitting when the primary interest is in relating genotypes and a trait. Finally, we consider composite input variables and discuss model interpretation.

*Nominal and ordinal predictors*

As described above, the tree-fitting methodology searches through the set of all predictor variables to identify the one that maximizes the reduction in node impurity. If a potential predictor $x$ is binary, then there is one possible split of individuals into the two daughter nodes based on the value of $x$. Specifically, those individuals with $x = 1$ go to one daughter node, say $\Omega_L$, and individuals for whom $x = 0$ go to the second node, $\Omega_R$. Now let us consider the setting in which $x$ is nominal, taking on the values $1, \ldots, m$. In this case, there are $m^* = \binom{m}{2} = m(m-1)/2$ ways of organizing individuals into two groups based on the value of $x$. For example, if $m = 3$, we have the following possible splits:

$$
\begin{aligned}
(1) \;\; i \in \begin{cases} \Omega_L & \text{if } x \in [1] \\ \Omega_R & \text{if } x \in [2,3] \end{cases} \\[2mm]
(2) \;\; i \in \begin{cases} \Omega_L & \text{if } x \in [1,2] \\ \Omega_R & \text{if } x \in [3] \end{cases} \\[2mm]
(3) \;\; i \in \begin{cases} \Omega_L & \text{if } x \in [1,3] \\ \Omega_R & \text{if } x \in [2] \end{cases}
\end{aligned}
\tag{6.14}
$$

Note that we do not distinguish between (1) above and

$$
(4) \;\; i \in \begin{cases} \Omega_L & \text{if } x \in [2,3] \\ \Omega_R & \text{if } x \in [1] \end{cases}
\tag{6.15}
$$

since the direction of the split is irrelevant. Thus, for $m = 3$ we have $m^* = 3$ possible splits given by Equation (6.14). We see that the number of possible splits, $m^*$, increases rapidly with $m$. The CART algorithms consider all of these possible splits for each of the potential predictors in choosing the best split. Notably, it is important in constructing a tree in R to specify whether a predictor variable is indeed nominal or in fact ordinal.

For an ordinal predictor $x$ that takes on $m$ levels, only $m - 1$ splits are considered. These are given by

$$
\begin{aligned}
(1) \;\; i \in \begin{cases} \Omega_L & \text{if } x \in [1] \\ \Omega_R & \text{if } x \in [2, \ldots, m] \end{cases} \\[2mm]
(2) \;\; i \in \begin{cases} \Omega_L & \text{if } x \in [1,2] \\ \Omega_R & \text{if } x \in [3, \ldots, m] \end{cases} \\[2mm]
\vdots \\[2mm]
(m-1) \;\; i \in \begin{cases} \Omega_L & \text{if } x \in [1, \ldots, m-1] \\ \Omega_R & \text{if } x \in [m] \end{cases}
\end{aligned}
\tag{6.16}
$$

The approach to continuous predictors is similar to that for ordinal variables. Consider for example that we want to include the continuous variable age as a potential predictor in our tree-fitting procedure. We begin by sorting the values of this variable from smallest to largest (or largest to smallest). All splits into two groups based on this ordered variable are then considered. In the following example, we illustrate generation of a regression tree for both ordinal and categorical inputs.

*Example 6.4 (Categorical and ordinal predictors in a tree).* In this example, we consider SNPs within the `resistin` gene and their association with the percentage change in non-dominant arm muscle strength using the FAMuSS data. Treating the SNPs as three-level categorical variables, we fit a regression tree as follows:

```
> attach(fms)
> Trait <- NDRM.CH
> library(rpart)
> RegTree <- rpart(Trait~resistin_c30t+resistin_c398t+
+             resistin_g540a+resistin_c980g+resistin_c180g+
+             resistin_a537c, method="anova")
> RegTree

n=611 (786 observations deleted due to missingness)

node), split, n, deviance, yval
      * denotes terminal node

1) root 611 665669.4 52.85352
  2) resistin_c980g=CC,CG 510 491113.4 51.23314 *
  3) resistin_c980g=GG 101 166455.3 61.03564 *
```

Based on this output, we see that `resistin_c980g` is the most predictive variable of `NDRM.CH`. Individuals who have the `CC` or `CG` genotypes have a predicted percentage change of 51.2, while individuals having the `GG` genotype have a predicted value of 61.0. This approach considers all splits of the three-level genotype variables into two groups. For example, for `resistin_980g`, we consider all three possible splits, given by

$$
(1) = \begin{cases} \Omega_L & \text{if } \mathtt{resistin\_c980g} \in (CC, CG) \\ \Omega_R & \text{if } \mathtt{resistin\_c980g} \in (GG) \end{cases}
$$

$$
(2) = \begin{cases} \Omega_L & \text{if } \mathtt{resistin\_c980g} \in (CC) \\ \Omega_R & \text{if } \mathtt{resistin\_c980g} \in (CG, GG) \end{cases} \qquad (6.17)
$$

$$
(3) = \begin{cases} \Omega_L & \text{if } \mathtt{resistin\_c980g} \in (CG) \\ \Omega_R & \text{if } \mathtt{resistin\_c980g} \in (CC, GG) \end{cases}
$$

Alternatively, we can treat genotype variables as ordinal. In this case, we only consider splits (1) and (2) above. In order to do this in our tree-fitting procedure, we define them as numeric variables as follows:

```
> RegTreeOr <- rpart(Trait~as.numeric(resistin_c30t)+
+            as.numeric(resistin_c398t)+as.numeric(resistin_g540a)+
+            as.numeric(resistin_c980g)+as.numeric(resistin_c180g)+
+            as.numeric(resistin_a537c), method="anova")
> RegTreeOr

n=611 (786 observations deleted due to missingness)

node), split, n, deviance, yval
      * denotes terminal node

1) root 611 665669.4 52.85352
  2) as.numeric(resistin_c980g)< 2.5 510 491113.4 51.23314 *
  3) as.numeric(resistin_c980g)>=2.5 101 166455.3 61.03564 *
```

In this case, the interpretation of the output is the same, though we see that now splits are defined according to whether the numeric version of resistin_c980g, which takes on the values 1, 2 and 3, is above or below a given threshold.                                                                                    □

The decision to include a variable as ordinal or nominal depends on prior knowledge about the effects of these variables on the trait. For example, suppose we are considering a quantitative trait $Y$ and a genotype that takes on the values $AA$, $AT$ and $TT$. There may be some prior scientific knowledge that suggests having one copy of the variant allele, $T$, cannot be more extreme than having two copies of $T$ in terms of increasing or decreasing $Y$. In this case, treating the genotype as an ordinal predictor, so that the possible nodes are defined by $(AA, AT)$ and $(TT)$ or $(AA)$ and $(AT, TT)$, makes the most sense. On the other hand, if no prior knowledge is available, we may also want to consider the split into nodes defined by $(AA, TT)$ and $(AT)$.

*Approaches to covariates*

There are several different approaches to handling clinical and demographic variables in the context of fitting a tree to data arising from a genetic association study. The first and simplest approach is just to ignore the covariates. That is, we can fit a tree to the trait based only on the genotype variables and without regard to the covariate information. This approach is similar to fitting an unadjusted regression model and in many instances can provide valuable information about disease etiology. The reasonableness of this approach, however, depends on the true underlying model structure, which is generally not known.

For example, suppose the covariate of interest is in the causal pathway between the genotypes under investigation and the disease outcome. One example of this might be an investigation of several SNPs and cardiovascular disease (CVD), where the measured covariate is body mass index (BMI). In this case, ignoring information about BMI may be reasonable since ultimately we are interested in identifying the genetic predictors of disease. On the other hand, if the covariate is an effect modifier, then ignoring it can result in a loss of power to detect true associations. For example, suppose the covariate is exposure to a specific drug, call it Chemical $X$. It is conceivable that the SNPs under investigation are predictive of CVD among individuals on Chemical $X$ but the genotype effects are negligible or even in the reverse direction among individuals that are not exposed to this chemical. In this case, ignoring information on the covariate may lead us to conclude that SNPs are not informative when in fact they are predictive within specific strata.

Alternatively, we can include the covariates along with the genotype variables as potential predictors in the model-fitting procedure. That is, suppose $\mathbf{Z} = (\mathbf{z}_1, \ldots, \mathbf{z}_m)$ represents the set of measured covariates, and let $\mathbf{X} = (\mathbf{x}_1, \ldots, \mathbf{x}_p)$ be genotype variables. This approach involves searching through the combined set $(\mathbf{Z}, \mathbf{X})$ to identify the best splitting variable at each node of the tree. This approach may offer an advantage over ignoring the covariates if in fact the covariates are effect modifiers or some form of conditional association exists. Returning to the example above, if exposure to Chemical $X$ is independently predictive of CVD, then the tree may first split on this variable. Subsequent splits are then conditional on exposure and the genotype variables would likely emerge within the exposed nodes. This approach is preferable if there is potential for complex structure, such as conditional associations, that is not known prior to model fitting.

Another approach is to stratify the sample prior to fitting the tree. If we know that the genotype effects are potentially different for certain levels of a factor variable, then this approach tends to be preferred. The advantage of a stratified analysis over including the factor variable as a potential predictor is that the variable does not need to be *more* predictive than the other predictor variables in order to be considered. Recall that the tree-splitting approach splits first on the *most* predictive variable. If our factor variable does not meet this requirement, then we may not discover these within-strata associations by simply including the variable as a potential predictor. On the other hand, if there is no prior knowledge about the structure of these conditional associations, then including the variables as predictors can help lead to these discoveries.

Finally, we can regress the trait on the covariates and then use the residuals from this model-fitting procedure as the outcome in fitting our tree. Again suppose our covariates are given by the design matrix $\mathbf{Z} = (\mathbf{z}_1, \ldots, \mathbf{z}_m)$ and we have a continuous trait $\mathbf{y}$. This approach begins by fitting the multivariable model

$$\mathbf{y} = \alpha + \mathbf{Z}\beta + \epsilon \tag{6.18}$$

Interaction terms among the covariates can also be included in the model above. Suppose the final model design matrix is given by $\mathbf{Z}^* = (\mathbf{z}_1, \ldots, \mathbf{z}_r)$ and the least squares estimate of the corresponding parameter vector is given by $\widehat{\beta}^*$. The fitted values of $\mathbf{y}$ from this model fit are then given by $\widehat{\mathbf{y}} = \mathbf{Z}^* \widehat{\beta}^*$ and the corresponding residuals are denoted $\mathbf{r} = \widehat{\mathbf{y}} - \mathbf{y}$. We now let $\tilde{\mathbf{y}} = \mathbf{r}$ and fit a tree with this new response variable and all of the genotype variables as predictors. The interpretation of the findings from this final approach will differ from those of the previous approaches. Specifically, in this case our tree includes the genotype variables that are predictive of the trait after taking out the effects of all of the covariates. That is, these variables capture information of the residual variability in $\mathbf{y}$ after accounting for the covariates. This approach may be optimal in the presence of confounding, though further research in this area is warranted.

*Composite predictors*

It is important to recognize that the tree structure allows discovery of a specific form of conditional association but is not well suited for discovering all types of associations. One subtle but important concept is that the tree algorithm is not searching specifically for statistical interaction; rather, classification and regression trees aim to identify conditional association. Consider for example a setting in which there is a relatively strong independent effect of a binary variable $\mathbf{x}_1$ on $\mathbf{y}$ such that the first split of our tree is on this variable. Now suppose that there is a statistical interaction between $\mathbf{x}_1$ and $\mathbf{x}_2$ so that the difference in the effect of $\mathbf{x}_2$ on $\mathbf{y}$ between $\mathbf{x}_1 = 1$ and $\mathbf{x}_1 = 0$ is given by the constant parameter $\gamma > 0$. Further suppose there is a predictor variable $\mathbf{x}_3$ that has the same effect on $\mathbf{y}$ for both levels of $\mathbf{x}_1$ and $\mathbf{x}_2$. Formally, we write this as a multivariable model given by

$$\mathbf{y} = \beta_0 + \beta_1 \mathbf{x}_1 + \beta_2 \mathbf{x}_2 + \beta_3 \mathbf{x}_3 + \gamma \mathbf{x}_1 * \mathbf{x}_2 + \epsilon \tag{6.19}$$

After our first split on $\mathbf{x}_1$, this model is expressed equivalently as:

$$\mathbf{y} = \begin{cases} \beta_0 + \beta_1 + (\beta_2 + \gamma)\mathbf{x}_2 + \beta_3 \mathbf{x}_3 + \epsilon & \text{for } \mathbf{x}_1 = 1 \\ \\ \beta_0 + \beta_2 \mathbf{x}_2 + \beta_3 \mathbf{x}_3 + \epsilon & \text{for } \mathbf{x}_1 = 0 \end{cases} \tag{6.20}$$

Now, suppose $\beta_3 > \beta_2 + \gamma$ and hence $\beta_3 > \beta_2$ since we assumed $\gamma$ is positive. In this case, the second split from both daughter nodes will likely be based on the value of $\mathbf{x}_3$ and not $\mathbf{x}_2$ even though only $\mathbf{x}_2$ interacts, in a statistical sense, with $\mathbf{x}_1$. That is, the conditional association is more relevant than the existence of a statistical interaction.

Another important aspect of the tree methodology is that in its original formulation it splits on a single variable. Thus, if polymorphisms at each of two genetic loci have an effect on the trait but neither SNP is independently predictive, the tree model may not uncover this association. One approach to addressing this limitation is to create composite predictor variables based on combinations of SNPs. For example, suppose again that there are two binary predictors, $\mathbf{x}_1$ and $\mathbf{x}_2$. One approach is to create the new nominal variable

$x_{(1,2)}$ that is simply a summary of the information in $x_1$ and $x_2$. For example, we can define

$$x_{(1,2)} = \begin{cases} 1 & \text{for } x_1 = 1, x_2 = 1 \\ 2 & \text{for } x_1 = 1, x_2 = 0 \\ 3 & \text{for } x_1 = 0, x_2 = 1 \\ 4 & \text{for } x_1 = 0, x_2 = 0 \end{cases} \tag{6.21}$$

If indeed there is statistical interaction between $x_1$ and $x_2$, then the tree-fitting algorithm is likely to split individuals into $\Omega_L = \{i : x_{(1,2),i} \in [1]\}$ and $\Omega_R = \{i : x_{(1,2),i} \in [2,3,4]\}$. Interestingly, this will occur whether or not there are main independent effects of $x_1$ and $x_2$. On the other hand, if there are simply additive effects of $x_1$ and $x_2$, then the first split will be $\Omega_L = \{i : x_{(1,2),i} \in [1,2]\}$ and $\Omega_R = \{i : x_{(1,2),i} \in [3,4]\}$. The second splits from both $\Omega_L$ and $\Omega_R$ will be $x_{(1,2)}$ as well. Alternative composite variables, such as may result from applying a clustering algorithm to the genotype data, are also tenable. This general approach is called patterning and recursive partitioning (PRP) and is closely related to multifactor dimensionality reduction (MDR) and the computational partitioning method (CPM). For additional details, see Foulkes *et al.* (2004), Ritchie *et al.* (2001) and Nelson *et al.* (2001).

## 6.2 Optimal trees

In the previous section, we focused on methods for building a tree. In general, the resulting tree needs to be *pruned* back into a smaller tree in order to apply more generally to another sample of data from the same population. As we include more splits in the tree, the resulting misclassification rate or mean square error will in general improve. However, at a certain point, these splits are only relevant to the sample of data under consideration and will not serve as an improvement in prediction for the population as a whole. This is similar to the overfitting problem in a regression setting. Recall that including additional variables in a regression will always improve the MSE; however, the model with all variables is generally not the optimal model. The best model is chosen based on the likelihood ratio test or a related procedure that accounts for the certainty that a variable is truly predictive.

For this reason, once we build a tree, it is important to determine the *optimal subtree*. This subtree will be a portion of the original tree and will provide the best fit to data arising from the general population from which our sample data were drawn. Extensive discussions of pruning techniques and the rationale behind different approaches is provided in Breiman *et al.* (1993) and Zhang and Singer (1999). Here we focus on two aspects of the pruning process. First, we discuss a method for arriving at *honest* estimates of error associated with a tree and then we describe *cost-complexity pruning*.

### 6.2.1 Honest estimates

Any tree we construct has an associated error that provides a measure of how well the tree predicts the observed data. Formally, the overall error associated with a tree $\mathcal{T}$, also referred to as *tree impurity*, is given by

$$R(\mathcal{T}) = \sum_{\tau \in \tilde{\mathcal{T}}} \pi(\tau) \times r(\tau) \qquad (6.22)$$

where $\tilde{\mathcal{T}}$ is the set of all terminal nodes in $\mathcal{T}$, $\pi(\tau)$ is the probability of belonging to the node given by $\tau$, and $r(\tau)$ is a corresponding measure of error for this node. In the classification setting, this within-node error is given by the misclassification rate, while in the regression setting, $r(\tau)$ is the mean square error. One estimate of this error, commonly referred to as the *resubstitution estimate*, is based simply on resubstituting the original data into the tree and calculating the corresponding quantities. As described above, inclusion of additional splits in a tree will always improve this resubstitution estimate of the error. In identifying SNPs associated with a disease trait, however, we want to ensure that an association is not simply a chance finding specific to the sample of data under consideration. To achieve this, we seek an *honest estimate* of the error, defined as one that applies to any sample taken from the population of interest.

Tenfold cross-validation (CV) is one approach that can be applied to arrive at an honest estimate of the error associated with a tree, $\mathcal{T}$. This involves dividing the individuals in our sample into ten approximately equal parts. Let these subsets be denoted $\mathcal{L}_i$ for $i = 1, \ldots, 10$. For each $i$, a tree is constructed based on the 9/10 of the data excluding $\mathcal{L}_i$, which we denote $\mathcal{L}_{-i}$ and call the *learning sample*. The error resulting from running the individuals in $\mathcal{L}_i$, called the *test sample*, through the tree is then recorded. Let the resulting error be denoted $R^{ts}(\mathcal{T}_{-i})$. An honest estimate of the error is then given by the average of the test sample errors over all $i = 1, \ldots, R$:

$$R^{CV}(\mathcal{T}) = \frac{1}{10} \sum_i R^{ts}(\mathcal{T}_{-i}) \qquad (6.23)$$

Note that this is an estimate of the error associated with the tree $\mathcal{T}$ constructed based on the entire sample, $\mathcal{L}_1 \cup \mathcal{L}_2 \cup \ldots \cup \mathcal{L}_{10}$.

### 6.2.2 Cost-complexity pruning

Choosing the *right-sized* tree is a critical aspect of applying the CART methodology. One approach to identifying such a tree involves first defining a measure of tree *cost complexity*, which takes into account both the error associated with the tree and the number of terminal nodes, called the size, of the tree. Formally, the cost complexity is given by

$$R_\alpha(\mathcal{T}) = R(\mathcal{T}) + \alpha|\mathcal{T}| \qquad (6.24)$$

where $R(\mathcal{T})$ is defined in Equation (6.22), $|\mathcal{T}|$ is the size of the tree $\mathcal{T}$ and $\alpha \geq 0$ is termed the *complexity parameter*. Here the idea is that we are adding a penalty for the number of nodes in the tree since additional nodes can make the tree more difficult to interpret. Identifying a right-sized tree is achieved through a pruning process that involves searching through subtrees to find the one that minimizes the cost complexity. Here a *subtree* of $\mathcal{T}$ is defined as a portion of $\mathcal{T}$ that excludes a given node and all offspring of that node. The search process is simplified by a theorem given by Breiman *et al.* (1993) that states that for a given value of the cost-complexity parameter, $\alpha$, there is a unique smallest subtree of $\mathcal{T}$ that minimizes $R_\alpha(\mathcal{T})$.

For a given internal node $\tau$, we define $R(\tau)$ to be the error associated with node $\tau$ and let $R(\mathcal{T}_\tau)$ equal the sum of errors associated with all terminal nodes that are offspring of $\tau$. There is a single value of $\alpha$ for which the corresponding cost complexities, $R_\alpha(\tau)$ and $R_\alpha(\mathcal{T}_\tau)$, for these two nested trees are equal. This is given by

$$\alpha_\tau = \frac{R(\tau) - R(\mathcal{T}_\tau)}{|\mathcal{T}_\tau| - 1} \qquad (6.25)$$

Notably, for $\alpha$ such that $\alpha \geq \alpha_\tau$, the subtree with terminal node $\tau$ is preferred to the tree that includes all offspring of $\tau$. In order to arrive at an optimal set of nested subtrees, we begin by letting $\alpha_{(1)}$ be the minimal $\alpha_\tau$ over all internal nodes $\tau$. This node is referred to as the *weakest link* in the tree. The first pruned subtree is then defined as the tree that converts all nodes $\tau$ to terminal nodes if $\alpha_\tau \leq \alpha_{(1)}$. It is in the application of a cross-validation procedure, as described below, that multiple nodes may meet this criterion. This subtree, denoted $\mathcal{T}_{\alpha_{(1)}}$, is now treated as the full tree, and $\alpha_\tau$ corresponding to each internal node is again recorded. That is, we recalculate the errors, $R(\tau)$ and $R(\mathcal{T}_\tau)$, for each internal node $\tau$ of $\mathcal{T}_{\alpha_{(1)}}$ and again solve for $\alpha_\tau$ using Equation (6.25). Now we define $\alpha_{(2)}$ as the smallest $\alpha_\tau$ over internal nodes of $\mathcal{T}_{\alpha_{(1)}}$ and define the second pruned subtree as above. This process is repeated until we arrive at a nested set of optimal subtrees given by

$$\mathcal{T}_{\alpha_{(m)}} \subset \mathcal{T}_{\alpha_{(m-1)}} \subset \ldots \subset \mathcal{T}_{\alpha_{(1)}} \subset \mathcal{T} \qquad (6.26)$$

and a corresponding sequence of complexity parameters

$$\alpha_{(m)} > \alpha_{(m-1)} > \ldots > \alpha_{(1)} > 0 \qquad (6.27)$$

The best subtree is the one that minimizes the true error associated with the tree. As described in Section 6.2.1, an honest estimate of this error is arrived at through application of a cross-validation procedure. Here we describe how to apply this procedure in the context of cost-complexity pruning. As in Section 6.2.1, we begin by dividing our sample into ten groups of approximately equal size and denote these $\mathcal{L}_i$ for $i = 1, \ldots, 10$. For each leaning sample, $\mathcal{L}_{-i}$, we determine the optimal set of subtrees based on the original complexity parameters, $\alpha_{(1)}, \ldots, \alpha_{(m)}$, that were derived using the entire set of data. An estimate of the error associated with each of these subtrees is then calculated using the test sample, $\mathcal{L}_i$. The average error over the $i = 1, \ldots, 10$ cross-validations is then determined for each $\alpha_{(k)}$ and denoted $R^{CV}(\mathcal{T}_{\alpha_{(k)}})$. This serves as an honest estimate of the error associated with the subtree $\mathcal{T}_{\alpha_{(k)}}$ from the full data. Finally, the last step accounts for the variability in the estimated error. Let $se_k$ denote the standard error of $R^{CV}(\mathcal{T}_{\alpha_{(k)}})$, and suppose $R^{CV}$ is minimized for $\alpha_{(k^*)}$. Since this error can be large, we select the best tree as the smallest tree such that the cross-validated estimate of error is within the interval

$$\left[ R^{CV}(\mathcal{T}_{\alpha_{(k^*)}}) - se_{k^*}, R^{CV}(\mathcal{T}_{\alpha_{(k^*)}}) + se_{k^*} \right] \tag{6.28}$$

An illustration of cost-complexity pruning is provided in the following example.

*Example 6.5 (Cost-complexity pruning).* In this example, we aim to characterize the association between mutations in the protease region of the viral genome and APV fold resistance based on the Virco data. Rather than dichotomizing the genotype data, as we did in Example 6.2, this time we use complete information on the observed amino acids at each site. We begin with the following code to generate a tree:

```
> attach(virco)
> VircoGeno <- data.frame(virco[,substr(names(virco),1,1)=="P"])
> library(rpart)
> Tree <- rpart(APV.Fold~.,data=VircoGeno)
> Tree

n=939 (127 observations deleted due to missingness)

node), split, n, deviance, yval
      * denotes terminal node

 1) root 939 356632.300 12.946540
   2) P54=-,A,L,MI,S,T,TI,TS,V,VA,VI,VIM,VL,X 889 237601.200 10.726550
     4) P46=-,ILM,IM,LM,LMI,MI,MIL,ML,V,VIM,X 481  44960.940   4.506653
       8) P54=-,T,TI,TS,VI,X 342    4475.893   1.980702 *
       9) P54=A,L,MI,S,V,VA 139  32934.020  10.721580
        18) P89=-,M,ML 132  24074.510   9.125000
          36) P82=-,A,AT,AV,S,T,TS 122  15185.570   7.560656 *
          37) P82=C,F,M,TA 10    4948.009  28.210000 *
        19) P89=I,V,VL 7    2178.014  40.828570 *
     5) P46=I,L,LI,LIM,VL 408 152093.900  18.059310
      10) P47=- 340 107235.300  15.332650
        20) P84=-,VI 232  56131.660  11.931900
          40) P50=-,L 214  37976.750  10.106540
```

```
        80) P33=-,I,LF,M,V 167   15681.720   7.290419 *
        81) P33=F,FL,IL,MIL 47   16264.770 20.112770 *
     41) P50=V 18     8964.740 33.633330 *
   21) P84=A,V,X 108   42656.790 22.637960
     42) P91=-,A,N,ST,Z 101   34293.240 20.867330
        84) P76=- 92   25472.030 18.966300 *
        85) P76=V 9   5090.080 40.300000 *
     43) P91=S,SA 7   3478.089 48.185710 *
   11) P47=A,V,VI 68   29691.790 31.692650
     22) P20=-,M,RK,T,TI,TK,VI 35    7891.207 23.125710 *
     23) P20=I,R,V 33   16507.440 40.778790
        46) P53=L,LF 13    4679.171 26.661540 *
        47) P53=- 20    7553.350 49.955000 *
 3) P54=LI,M,MIL,VM 50   36749.950 52.418000
   6) P20=-,M,R,TK,V 33   21075.000 43.024240
     12) P46=- 11    5723.236 20.881820 *
     13) P46=I,IM,V,VI 22    7262.030 54.095450 *
   7) P20=I,QK,RK,T,VI 17    7110.222 70.652940 *
```

Pruning this tree involves determining a cost-complexity parameter value. We do this by considering the output of the printcp() function. A visual representation is given by applying the plotcp() function similarly and is illustrated in Figure 6.3:

```
> plotcp(Tree)
> printcp(Tree)

Regression tree:
rpart(formula = APV.Fold ~ ., data = VircoGeno)

Variables actually used in tree construction:
 [1] P20 P33 P46 P47 P50 P53 P54 P76 P82 P84 P89 P91

Root node error: 356632/939 = 379.8

n=939 (127 observations deleted due to missingness)
```

|    | CP | nsplit | rel error | xerror | xstd |
|----|----|--------|-----------|--------|------|
| 1  | 0.230717 | 0  | 1.00000 | 1.00245 | 0.080943 |
| 2  | 0.113693 | 1  | 0.76928 | 0.80572 | 0.066653 |
| 3  | 0.042528 | 2  | 0.65559 | 0.69204 | 0.058295 |
| 4  | 0.024727 | 3  | 0.61306 | 0.68130 | 0.057954 |
| 5  | 0.024016 | 5  | 0.56361 | 0.69880 | 0.060721 |
| 6  | 0.022684 | 6  | 0.53959 | 0.69188 | 0.061341 |
| 7  | 0.021173 | 7  | 0.51691 | 0.69536 | 0.061128 |
| 8  | 0.018735 | 8  | 0.49574 | 0.68929 | 0.061072 |
| 9  | 0.016909 | 9  | 0.47700 | 0.67763 | 0.060596 |
| 10 | 0.014842 | 10 | 0.46009 | 0.66358 | 0.058020 |
| 11 | 0.013699 | 11 | 0.44525 | 0.66136 | 0.058263 |
| 12 | 0.011987 | 12 | 0.43155 | 0.66654 | 0.058662 |

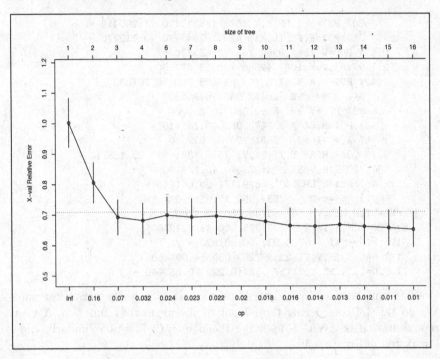

**Fig. 6.3.** Cost-complexity pruning for Example 6.5

```
13 0.011050      13    0.41956 0.66042 0.058219
14 0.010462      14    0.40851 0.65628 0.058176
15 0.010000      15    0.39805 0.65099 0.057866
```

Based on the output and figure, we see that the cross-validated error begins to increase between three and five splits, corresponding to four and six terminal nodes. We therefore want to select the tree with four terminal nodes. This is achieved using the **prune()** function and specifying, in this example, **cp=0.03** as follows:

```
> pruneTree <- prune(Tree,cp=.03)
> pruneTree

n=939 (127 observations deleted due to missingness)

node), split, n, deviance, yval
      * denotes terminal node

  1) root 939 356632.30 12.946540
    2) P54=-,A,L,MI,S,T,TI,TS,V,VA,VI,VIM,VL,X 889 237601.20 10.726550
      4) P46=-,ILM,IM,LM,LMI,MI,MIL,ML,V,VIM,X 481  44960.94  4.506653 *
      5) P46=I,L,LI,LIM,VL 408 152093.90 18.059310
        10) P47=- 340 107235.30 15.332650 *
```

```
   11) P47=A,V,VI 68   29691.79 31.692650 *
  3) P54=LI,M,MIL,VM 50   36749.95 52.418000 *
```

In this example, we see that many of the splits in the original tree are chance findings. Application of cost-complexity pruning allows us to arrive at a tree that is likely to reduce error in an independent test sample.     □

# Problems

**6.1.** Describe in words the motivation for cost-complexity pruning.

**6.2.** Using the Virco data, construct and prune a classification tree to determine whether any mutations within the protease region of the viral genome are associated with greater APV fold resistance than IDV fold resistance. Here define the trait as an indicator variable for `APV.fold > IDV.fold`. Explain your findings.

**6.3.** Using the Virco data, construct and prune a regression tree to determine whether any mutations within the protease region of the viral genome are associated with a difference in APV and IDV fold resistance. Here define the trait as `APV.fold-IDV.fold`. Explain your findings.

**6.4.** Describe three different approaches to handling covariates in fitting a classification or regression tree. Give a setting in which one approach may be preferable.

**6.5.** Fit a regression tree to the FAMuSS data using SNPs within the `actn3` and `resistin` genes as potential predictor variables and change in non-dominant arm muscle strength, as measured by `NDRM.CH`, as the dependent variable. Fit the tree with and without `Race` as a potential predictor in the model. Interpret your findings.

# Additional Topics in High-Dimensional Data Analysis

In this chapter, we describe several additional approaches that are well suited to high-dimensional data settings. Three of these, namely random forests (RFs) (Section 7.1), logic regression (Section 7.2) and multivariable adaptive regression splines (MARS) (Section 7.3), extend the tree framework outlined in Chapter 6. Random forests were originally proposed by Breiman (2001), and logic regression was first described by Kooperberg *et al.* (2001), Ruczinski *et al.* (2003) and Ruczinski *et al.* (2004). A complete description of the MARS methodology can be found in Friedman (1991). We also present a brief description of Bayesian variable selection (Section 7.4), with particular emphasis on fundamental concepts that will guide the reader in further explorations. Additional readings on Bayesian variable selection methods and related extensions include George and McCulloch (1993), Chipman *et al.* (1998), Brown *et al.* (2002), West (2003), Lunn *et al.* (2006) and Hoggart *et al.* (2008), among others. Bayesian variable selection approaches are becoming increasingly popular in the analysis of high-throughput genotype data for identifying sets of SNPs that are associated with the trait under investigation. Finally, we end with a listing of several alternative high-dimensional data tools and their sources that may provide additional insight in characterizing genotype–trait associations (Section 7.5).

Random forests are comprised of an ensemble of trees, and with their application we see a notable paradigm shift from *characterizing the structure* of association to *quantifying the importance* of variables. Logic regression, on the other hand, provides us with clear rules of association similar to CART, though these can be more complex and challenging to interpret biologically. Use of variable importance scores in logic regression, as described in Kooperberg and Ruczinski (2005) and Schwender and Ickstadt (2008a) and in the original formulation of random forests, may similarly serve as a useful application of this approach. In this chapter, we focus on the application of these methods as an exploratory tool for identifying SNP–trait associations warranting further investigations. Assessing the statistical significance of variable importance measures arising from machine learning algorithms has

A.S. Foulkes, *Applied Statistical Genetics with R: For Population-based Association Studies*, Use R, DOI: 10.1007/978-0-387-89554-3_7, © Springer Science+Business Media LLC 2009 received recent attention in the literature, and the advanced reader is referred to van der Laan (2006) for a discussion. Multivariable adaptive regression splines (MARS) involve a recursive partitioning algorithm similar to the one we saw in Chapter 6 while offering some additional advantages in the context of alternative models of association. This approach was originally formulated in the context of continuous predictor variables but performs reasonably well for categorical inputs. As we will see below, the output of MARS includes information on both the set of variables that explain variability in the trait and the nature of their association. Finally, while there are several advanced Bayesian variable selection approaches, the methodology presented in this chapter provides for determining a group of potentially relevant predictor variables, with less emphasis on the precise structure of association.

The methods presented herein are broadly referred to as machine learning approaches, as they involve computationally intensive algorithms for mining high-dimensional data. The choice of methods is based on their intuitive appeal, their solid analytic grounding and their growing popularity in the scientific literature for the analysis of SNP–trait associations. This presentation is not intended to provide a comprehensive overview of data-mining methods, as offered, for example, in Hastie *et al.* (2001). Instead, we describe a few key approaches and their specific applications to population-based association studies as a means of introducing the reader to the inherent value of more sophisticated, non-standard approaches to analysis.

## 7.1 Random forests

Random forests (RFs) represent an extension of CART that involves generating an ensemble of classification or regression trees. This set of trees is constructed in a manner that addresses some of the limitations of CART. Most notably, through repeated sampling of sets of predictor variables at the tree-splitting stage, RFs offer a natural approach to handling collinearity among SNPs. Unlike in the CART setting, RFs do not yield a clear structure for the final model of association; i.e., we do not generate a final tree that can be interpreted as a model for the genotype–trait association. Instead, a measure of variable importance resulting from application of the approach provides us with a general measure of the contribution of each potential predictor variable to the observed variability in the trait under investigation. Further details on this approach can be found in Breiman (2001), with an application to human genotype data given in Bureau *et al.* (2005) and an application to HIV genetic data provided in Segal *et al.* (2004). In this section, we focus primarily on methods for generating a random forest and variable importance scores (Section 7.1.1) as well as missing data considerations (Section 7.1.2). A brief discussion of how to handle covariates is also provided (Section 7.1.3), though the reader is referred to Section 6.1.3 for a more thorough discussion of how to handle varying types of inputs.

### 7.1.1 Variable importance

We begin by reviewing our notation. Let $\mathbf{X} = (\mathbf{x}_1, \ldots, \mathbf{x}_p)$ be the set of $p$ potential predictor variables, where $\mathbf{x}_p = (x_{1p}, \ldots, x_{np})^T$, and suppose $\mathbf{y}$ is the trait under investigation, where $n$ is the number of individuals in our sample. Recall that in Chapter 6 we defined several measures of node impurity that describe the deviations of the observed data from the predicted value within a node, where a node is defined based on the observed values of $\mathbf{X}$ (see Equations (6.7), (6.10) and (6.13)). For example, in the context of a quantitative trait, the most common measure of node impurity is given by the mean square error (MSE), formalized for a node $\Omega$ by

$$I(\Omega) = \frac{1}{n_\Omega} \sum_{i \in \Omega} (y_i - \bar{y})^2 \tag{7.1}$$

where $n_\Omega$ is the number of individuals in node $\Omega$ and $\bar{y}$ is the mean of the corresponding elements of $\mathbf{y}$. In this setting, the *best* split is the one that maximizes the reduction in node impurity, given by

$$\phi = I(\Omega) - I(\Omega_L) - I(\Omega_R) \tag{7.2}$$

where $\Omega_L$ and $\Omega_R$ are the left and right daughter nodes of $\Omega$, respectively.

A step-by-step summary of the RF algorithm is given in Algorithm 7.1 below. In the first step of the algorithm, we randomly select approximately two-thirds of our sample. This is called the *learning sample (LS)* since it is used to grow an initial tree. The remaining individuals constitute the *out-of-bag (OOB) data* and are used to evaluate how well the tree applies to data that were not used to generate it. Note that the concept of OOB data is similar to test-sample data. In the second step of the algorithm, we grow a tree based on the LS data. Here the methodology of Chapter 6 is applied, with two notable departures. First, in the context of RFs, we fit an *unpruned* tree. Recall that pruning is an important aspect of the tree-fitting methodology that ensures applicability of the findings to alternative samples from the same population, as presented in Section 6.2. Evaluating the relative importance of each predictor variable in independent test samples is instead incorporated in step 3 of the algorithm.

The second notable difference in fitting the tree at this stage is that for each node only a subset of the variables are considered as potential predictors. That is, instead of determining the best split among all potential predictors, a random sample of these variables (typically about a third of the variables) are considered as potential splitting variables. A primary advantage of drawing a random subset of potential predictor variables at each node is that it offers a natural approach to handling collinearities in the data. Consider the case in which two SNPs are in high linkage disequilibrium. If a tree splits on one of these two SNPs, then there may be insufficient variability in the second SNP among individuals in the resulting daughter nodes to split additionally on this SNP. Consideration of *surrogate splits* in the usual CART framework allows us to investigate this phenomenon. In the forest setting, by taking a subset of the SNPs, the expectation is that in some instances the first of two correlated SNPs will be chosen as a potential predictor, while in other instances the second SNP will be selected. In theory, the resulting variable importance scores will then reflect the contribution of each SNP.

The next step of the algorithm involves using the OOB data to evaluate the importance of each potential predictor variable. First, the overall *tree impurity* is measured by running the OOB data through the tree. Tree impurity is defined simply as the sum of impurity measures over all terminal nodes of the tree. *Variable importance* is calculated for each potential predictor as the difference between this impurity and the overall tree impurity when the corresponding predictor variables are permuted. Conceptually, the idea here is that permuting the value of an unimportant variable will lead to no change in the overall tree impurity, while permuting an influential variable will lead to an increase in the overall impurity measure. Finally, as noted in steps 4 and 5 of the algorithm, each of the steps just described are repeated multiple times and the average importance scores are recorded for each variable.

---

ALGORITHM 7.1: RANDOM FORESTS: We begin by initializing $b = 1$.

1. Randomly sample with replacement $n_1$ (approximately equal to $2n/3$) individuals, and call this the learning sample (LS). Let the remaining $n_2 = n - n_1$ individuals represent the out-of-bag (OOB) data.

2. Using the LS data only, generate an unpruned tree by randomly sampling a subset of the $p$ predictors at each node to be considered as potential splitting variables.

3. Based on the OOB data only:
   a) Record the overall tree impurity, and let this be denoted $\pi_b$.
   b) Permute $\mathbf{x}_j$, and record overall tree impurity using the permuted data for each $j = 1, \ldots, p$. Denote this $\pi_{bj}$ and define the variable importance for the $j$th predictor as $\delta_{bj} = \pi_{bj} - \pi_b$.

4. Repeat steps (1)–(3) for $b = 2, \ldots, B$ to obtain $\delta_{1j}, \ldots, \delta_{Bj}$ for each $j$.

5. Record the overall variable importance score for $\mathbf{x}_1, \ldots, \mathbf{x}_p$, defined for the $j$th predictor as

$$\widehat{\theta}_j = \frac{1}{B} \sum_{b=1}^{B} \delta_{bj} \tag{7.3}$$

---

Fitting a random forest in R is straightforward using the `randomForest` package, as following example illustrates. This package provides a standardized measure of variable importance, given by $\widehat{\theta}_j/SE(\widehat{\theta}_j)$, where $\widehat{\theta}_j$ is defined in Equation (7.3) and $SE(\widehat{\theta}_j)$ is the standard deviation of $\delta_{bj}$ over the $B$ trees divided by the square root of $B$. This measure is denoted %IncMSE in the R output and referred to as the mean decrease in accuracy. Also provided in this output is the average over the $B$ trees of the total node impurity explained by splits based on the corresponding variable. This is also calculated based on the OOB data but does not involve permutations of the variables. This latter measure is labeled IncNodePurity in the R output and referred to as the mean decrease in node impurity.

*Example 7.1 (An application of random forests).* In this example, we apply the RF approach to the Virco data, as described in Example 6.3 using the R function randomForest() in the randomForest package. The quantitative trait is again given by the difference in NFV and IDV fold resistance, and the potential predictor variables are indicators for a mutation at each of the 99 amino acids in the protease region. Recall that application of the regression tree approach in Example 6.3 yielded a tree with splits on the variables P35, P46, P54, P58 and P73 prior to pruning. We begin by installing and loading the randomForest package using the following commands:

```
> install.packages("randomForest")
> library(randomForest)
```

We then define the trait and a matrix of potential predictor variables, given by Trait and VircoGeno, respectively, using the same code as in Example 6.3. The randomForest() function does not permit missing data in the response variable, and so we next subset our data to include only those individuals with complete information on the trait:

```
> attach(virco)
> Trait <- NFV.Fold - IDV.Fold
> VircoGeno <- data.frame(virco[,substr(names(virco),1,1)=="P"]!="-")
> Trait.c <- Trait[!is.na(Trait)]
> VircoGeno.c <- VircoGeno[!is.na(Trait),]
```

Finally, we fit a random forest and plot the ordered variable importance measures, given by mean decrease in accuracy (%IncMSE) and mean decrease in node impurity (%IncMSE), using the randomForest() and varImpPlot() functions. Note that your output may vary slightly since the RF approach involves randomly selecting individuals to constitute the LS data and OOB data for each tree and randomly sampling the predictors for consideration at each split of a tree.

```
> RegRF <- randomForest(VircoGeno.c, Trait.c, importance=TRUE)
> RegRF

Call:
 randomForest(x = VircoGeno.c, y = Trait.c, importance = TRUE)
```

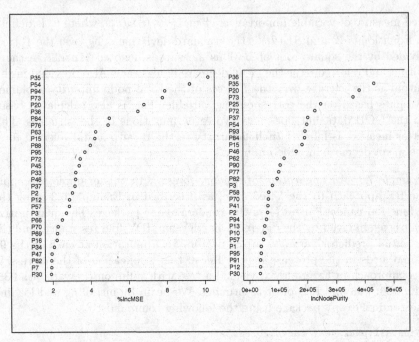

**Fig. 7.1.** Ordered variable importance scores from random forest

```
              Type of random forest: regression
                   Number of trees: 500
No. of variables tried at each split: 33

         Mean of squared residuals: 5745.777
                   % Var explained: 12.89
```

```
> varImpPlot(RegRF)
```

The resulting plot is given in Figure 7.1. Here we see that the five most important sites, as measured by the mean decrease in accuracy, are P20, P35, P54, P73 and P94. Interestingly, P20 and P94 were not identified using the regression tree approach of Example 6.3. It is also important to emphasize here that the top-ranked variables are not necessarily statistically significant predictors of the trait; rather, they represent genotype indicators that are *most* predictive among those considered and are thus worthy of further investigation. Application of the `getTree()` allows a user to further investigate the structure of the trees with an RF. However, remember that, depending on which tree number is specified (given by k in the function call), the results can vary dramatically, and so interpretation of this structure is tenuous.    □

## 7.1.2 Missing data methods

In the example above, the genotype data are fully observed, though in a more general setting we expect to observe some missingness in the predictor variables. Here we again distinguish between two general types of missing data in the context of population genetic association studies: (1) data that were intended to be collected but are missing and (2) unobservable data arising as a result of ambiguity in allelic phase.

*Missing genotype data*

We begin by considering the setting in which we have missingness in the SNP data with complete information on the measured trait. There are several simple yet reasonable approaches that can be taken in this setting. If the missingness is truly random, we can simply remove individuals from our analysis who have missing genotype data. Practically, however, this is not always feasible since in the context of a large number of genotype variables it is common for the majority of individuals to have missingness in at least one of these variables.

Another approach is to replace each missing value with the genotype that is most frequent in our sample for the corresponding SNP. Formally, we let $x_{ij}$ represent the genotype for individual $i$ at SNP $j$ and suppose $x_{ij}$ is missing. Suppose further that the observed alleles at site $j$ are $g_{j1} = AA$, $g_{j2} = Aa$ and $g_{j3} = aa$, with sample proportions of $\widehat{\pi}_{j1}$, $\widehat{\pi}_{j2}$ and $\widehat{\pi}_{j3}$, respectively. Single imputation involves replacing the missing value for $x_{ij}$ with the genotype that is most common at this site. That is, we let

$$x_{ij} = \sum_{k=1}^{3} g_{jk} \times I\left[\max\left(\widehat{\pi}_j\right) = \widehat{\pi}_{jk}\right] \qquad (7.4)$$

where $I[\cdot]$ is the indicator function taking on the value of 1 if its argument is true and 0 otherwise, and $\max\left(\widehat{\pi}_j\right)$ is the maximum over the set $(\widehat{\pi}_{j1}, \widehat{\pi}_{j2}, \widehat{\pi}_{j3})$, where we assume the three estimated proportions are not equal. Genotype data are similarly substituted for all $i$ and $j$ for which $x_{ij}$ is missing. This step yields a single completed dataframe, as illustrated in the following example using the `na.roughfix()` function within the `randomForest` package. Importantly, imputing genotypes should be applied within each racial and ethnic group separately since genotype frequencies tend to differ by these self-declared groupings.

*Example 7.2 (RF with missing SNP data—single imputation).* Suppose we are interested in characterizing association between 24 `AKT1` SNPs, for which we expect some missing data, and the change in the non-dominant arm muscle strength before and after exercise training, using the FAMuSS data. For illustration, we focus only on Caucasians and begin by defining our trait variable and genotype matrix:

```
> attach(fms)
> Trait <- NDRM.CH[Race=="Caucasian" & !is.na(Race) & !is.na(NDRM.CH)]
> NamesAkt1Snps <- names(fms)[substr(names(fms), 1, 4)=="akt1"]
> FMSgeno <- fms[,is.element(names(fms), NamesAkt1Snps)][
+       Race=="Caucasian" & !is.na(Race) & !is.na(NDRM.CH),]
```

Note that in the code above subjects for whom the corresponding trait information is missing were also removed. This step is necessary since the current implementation of random forests in R does not allow missing data on the outcome. The dimensions of the genotype matrix tell us how many individuals and SNPs remain under consideration:

```
> dim(FMSgeno)
```

```
[1] 777   24
```

Recall that the first element in the output above is the number of rows of the input matrix, in our case the number of individuals, $n = 777$, and the second element is the number of columns, in our case the number of SNPs, $p = 24$.

Using the apply() and is.na() functions, we see that there is 2–8% missingness across the 24 akt1 SNPs:

```
> round(apply(is.na(FMSgeno), 2, sum)/dim(FMSgeno)[1],3)
```

| akt1_t22932c | akt1_g15129a | akt1_g14803t |
|---|---|---|
| 0.069 | 0.067 | 0.021 |
| akt1_c10744t_c12886t | akt1_t10726c_t12868c | akt1_t10598a_t12740a |
| 0.071 | 0.075 | 0.071 |
| akt1_c9756a_c11898t | akt1_t8407g | akt1_a7699g |
| 0.066 | 0.069 | 0.067 |
| akt1_c6148t_c8290t | akt1_c6024t_c8166t | akt1_c5854t_c7996t |
| 0.021 | 0.066 | 0.021 |
| akt1_c832g_c3359g | akt1_g288c | akt1_g1780a_g363a |
| 0.021 | 0.021 | 0.021 |
| akt1_g2347t_g205t | akt1_g2375a_g233a | akt1_g4362c |
| 0.066 | 0.066 | 0.069 |
| akt1_c15676t | akt1_a15756t | akt1_g20703a |
| 0.021 | 0.021 | 0.021 |
| akt1_g22187a | akt1_a22889g | akt1_g23477a |
| 0.071 | 0.021 | 0.021 |

One approach to handling these missing data in the predictor variables is single imputation using the na.roughfix() function of the randomForest package as follows:

```
> library(randomForest)
> FMSgenoRough <- na.roughfix(FMSgeno)
```

In the case of factor variables, this function replaces missing data with the most frequent value of the corresponding variable. For example, consider the distribution of genotypes for the first SNP, given by

```
> table(FMSgeno$"akt1_t22932c")

CC  TC  TT
 3  55 665
```

Since the TT genotype has the highest frequency, the `na.roughfix()` function replaces all missing values for `akt1_t22932c` with TT. The resulting output has no missing data, as seen below:

```
> round(apply(is.na(FMSgenoRough), 2, sum)/dim(FMSgeno)[1],3)
```

| akt1_t22932c | akt1_g15129a | akt1_g14803t |
|---|---|---|
| 0 | 0 | 0 |
| akt1_c10744t_c12886t | akt1_t10726c_t12868c | akt1_t10598a_t12740a |
| 0 | 0 | 0 |
| akt1_c9756a_c11898t | akt1_t8407g | akt1_a7699g |
| 0 | 0 | 0 |
| akt1_c6148t_c8290t | akt1_c6024t_c8166t | akt1_c5854t_c7996t |
| 0 | 0 | 0 |
| akt1_c832g_c3359g | akt1_g288c | akt1_g1780a_g363a |
| 0 | 0 | 0 |
| akt1_g2347t_g205t | akt1_g2375a_g233a | akt1_g4362c |
| 0 | 0 | 0 |
| akt1_c15676t | akt1_a15756t | akt1_g20703a |
| 0 | 0 | 0 |
| akt1_g22187a | akt1_a22889g | akt1_g23477a |
| 0 | 0 | 0 |

Finally, we fit a random forest to the imputed data and print the resulting importance scores:

```
> RandForRough <- randomForest(FMSgenoRough, Trait, importance=TRUE)
> RandForRough$"importance"[order(RandForRough$"importance"[,1],
+      decreasing=TRUE),]
```

|  | %IncMSE | IncNodePurity |
|---|---|---|
| akt1_t10726c_t12868c | 230.835721 | 25870.501 |
| akt1_t8407g | 222.902558 | 17419.779 |
| akt1_g14803t | 131.418231 | 18502.134 |
| akt1_g288c | 131.128453 | 23422.797 |
| akt1_a15756t | 111.510235 | 24609.514 |
| akt1_c6148t_c8290t | 109.743902 | 17091.662 |
| akt1_g15129a | 105.979096 | 14961.767 |
| akt1_c832g_c3359g | 97.385668 | 11066.443 |
| akt1_a22889g | 92.513700 | 27341.961 |
| akt1_g2347t_g205t | 87.862083 | 17249.624 |
| akt1_t10598a_t12740a | 87.697216 | 18646.466 |
| akt1_g22187a | 74.891706 | 22527.999 |
| akt1_c6024t_c8166t | 66.268968 | 9055.988 |
| akt1_c5854t_c7996t | 57.670909 | 18339.356 |
| akt1_a7699g | 57.413211 | 7927.382 |

| | | |
|---|---|---|
| akt1_c9756a_c11898t | 49.213650 | 7750.871 |
| akt1_g23477a | 49.025595 | 20268.004 |
| akt1_g2375a_g233a | 48.220095 | 16762.174 |
| akt1_c15676t | 42.744631 | 21519.880 |
| akt1_g4362c | 34.600182 | 8820.092 |
| akt1_g1780a_g363a | 6.348848 | 3298.015 |
| akt1_g20703a | 5.970387 | 29753.180 |
| akt1_c10744t_c12886t | 5.111987 | 4074.560 |
| akt1_t22932c | -0.903492 | 17343.908 |

In the code above, we use the `order()` function to order the importance scores by the `%IncMSE`, which is represented by the first column of the `RandForRough$"importance"` matrix in our example. Again, repeated application of this approach will likely yield slightly different results since the RF algorithm involves random sampling.                                     □

As noted above, this imputation approach should be applied within racial and ethnic strata since the corresponding genotype frequencies can vary substantially. However, once an imputed genotype matrix is generated within each racial and ethnic group, as given by `FMSgenoRough` for Caucasians in the example above, the data can be recombined and subsequent analysis of association can be performed on the combined data. That is, we can impute genotypes within each racial and ethnic group and then combine the data prior to fitting a random forest.

An alternative, more sophisticated approach to handling missing categorical predictor variables is described in Breiman (2003) and implemented with the `rfImpute()` function in the `randomForest` package. A primary difference between this approach and the single-imputation approach described above is that the approach of Breiman (2003) uses information about the trait to inform the SNP reconstruction. Recall that the single-imputation approach uses only the observed genotype information to assign a value to missing genotype data. Secondly, the Breiman (2003) approach involves repeatedly updating the missing data. A step-by-step summary of this multiple imputation approach is given in ALGORITHM 7.2 below.

The first step of this algorithm is the single-imputation approach described above and illustrated in Example 7.2. The second step involves first fitting a random forest using ALGORITHM 7.1 based on these imputed data and subsequently determining proximity scores between all pairs of individuals. The *proximity score* between any two individuals is defined as the proportion of trees within a forest in which the corresponding pair of subjects falls into the same terminal node. If there are $n$ individuals in our sample, then the *proximity matrix* is the $n \times n$ matrix of proximity scores between all pairs of individuals. By definition, the diagonal elements of the proximity matrix are equal to 1.

The third step of this approach is to replace each missing value for each individual with the value of the corresponding variable with the highest average proximity score. For example, suppose that after the first imputation the proximity matrix is given by

$$\mathbf{P} = \begin{pmatrix} p_{11} & p_{12} & \cdots & p_{1n} \\ p_{21} & p_{22} & \cdots & p_{2n} \\ & \vdots & \\ p_{n1} & p_{n2} & \cdots & p_{nn} \end{pmatrix} \tag{7.5}$$

Further suppose $x_{ij}$, the genotype at the $j$th SNP for individual $i$, is missing. At this step, we determine, within each observed level of this SNP, the average proximity scores between individual $i$ and all other individuals. For example, suppose the observed genotypes for SNP $j$ are $g_{j1}$, $g_{j2}$ and $g_{g3}$. Then, for $k = 1, 2, 3$, we calculate

$$\bar{p}_k = \frac{1}{n_k} \sum_{l(l \neq i)} p_{il} \times I[x_{lj} = g_{jk}] \tag{7.6}$$

where $n_k$ is the number of individuals with genotype $g_{jk}$ and $I[\cdot]$ is again the indicator function. Here we are simply summing the proximity scores between individual $i$ and all other individuals that have genotype $g_{jk}$ and then dividing by the number of such individuals. The missing data, $x_{ij}$, are then replaced by the genotype level, $g_{jk}$, with the maximum corresponding value of $\bar{p}_k$.

In the context of a continuous predictor, a weighted average of the non-missing values is used in place of the missing data, where the weights are equal to the proximity scores. The second and third steps of this algorithm are repeated a prespecified number of times. Notably, a convergence criterion is not applied to determine how many repetitions of these steps are necessary for stability. Finally, a random forest is fit to the final set of imputed data. This multiple-imputation approach is illustrated in the following example.

---

ALGORITHM 7.2: RF WITH MISSING SNP DATA—MULTIPLE IMPUTATION

1. Replace each missing genotype variable with the value of this variable that is most frequently observed within the sample; i.e., across all individuals.

2. Fit a random forest based on ALGORITHM 7.1 and determine proximity scores for each pair of individuals.

3. Assign each missing value the value of the corresponding variable with the highest average proximity score.

4. Repeat steps (2) and (3) multiple times.

5. Fit a random forest to the final imputed dataset.

*Example 7.3 (RF with missing SNP data—multiple imputation).* Returning to
Example 7.2, suppose we are again interested in characterizing the association
between the `akt1` SNPs and change in the non-dominant arm muscle strength
based on the FAMuSS data. We begin by defining our trait variable and
genotype matrix, excluding all observations with missing trait information:

```
> attach(fms)
> Trait <- NDRM.CH[Race=="Caucasian" & !is.na(Race) & !is.na(NDRM.CH)]
> NamesAkt1Snps <- names(fms)[substr(names(fms),1,4)=="akt1"]
> FMSgeno <- fms[,is.element(names(fms), NamesAkt1Snps)][
+       Race=="Caucasian" & !is.na(Race) & !is.na(NDRM.CH),]
```

Next we multiply impute the genotype data as follows. By default, `iter=5`
imputations are implemented and `ntree=300` trees are generated per iteration.

```
> library(randomForest)
> FMSgenoMI <- rfImpute(FMSgeno,Trait)
```

```
        |        Out-of-bag  |
Tree  |        MSE   %Var(y) |
300   |       1253   110.44  |
        |        Out-of-bag  |
Tree  |        MSE   %Var(y) |
300   |       1236   108.96  |
        |        Out-of-bag  |
Tree  |        MSE   %Var(y) |
300   |       1224   107.86  |
        |        Out-of-bag  |
Tree  |        MSE   %Var(y) |
300   |       1227   108.15  |
        |        Out-of-bag  |
Tree  |        MSE   %Var(y) |
300   |       1230   108.42  |
```

Finally, we fit a random forest with the final imputed genotype data and print
the corresponding ordered importance scores. The first column of the output
of `rfImpute()` is the trait and is therefore removed in specifying the genotype
data for the `randomForest()` function:

```
> RandForFinal <- randomForest(FMSgenoMI[,-1], Trait, importance=TRUE)
> RandForFinal$"importance"[order(RandForFinal$"importance"[,1],
+       decreasing=TRUE),]
```

```
                        %IncMSE  IncNodePurity
akt1_t10726c_t12868c  307.145703     26114.672
akt1_t8407g           265.213630     19725.956
akt1_g288c            121.367971     22595.940
akt1_a15756t          119.265290     25032.467
akt1_g15129a          118.072551     14539.725
akt1_c6148t_c8290t    112.527852     14320.644
```

```
akt1_t10598a_t12740a  111.086340      16453.273
akt1_g14803t          109.884890      17827.457
akt1_a22889g          107.288643      29130.354
akt1_c832g_c3359g      91.884698      12015.553
akt1_g2347t_g205t      91.641632      17300.452
akt1_g22187a           77.053093      23320.635
akt1_g2375a_g233a      71.775880      21854.456
akt1_a7699g            70.866588       9456.752
akt1_c6024t_c8166t     65.666912       8284.760
akt1_c9756a_c11898t    54.905616       8634.346
akt1_g23477a           53.646350      19866.228
akt1_c5854t_c7996t     52.970776      19684.519
akt1_g4362c            45.275224      10084.659
akt1_c15676t           38.993791      21298.354
akt1_c10744t_c12886t   22.439268      12000.891
akt1_g1780a_g363a      10.307277       3639.588
akt1_g20703a            2.991968      28032.704
akt1_t22932c           -1.287238      18345.782
```

Interestingly, the SNP with the highest importance score is given by t10726c_t12868c using this approach as well as the single-imputation approach of Example 7.2; however, the corresponding %IncMSE is greater in this example. This may be a result of the difference in the completed data resulting from the two algorithms. This difference is highlighted by creating a $2 \times 2$ table using the table() function as follows:

```
> table(FMSgenoMI$akt1_t10726c_t12868c,
+        FMSgenoRough$akt1_t10726c_t12868c)

      CC  TC  TT
  CC 599   0   0
  TC   6 139   0
  TT  24   0   9
```

From this output, we see that $n = 24$ subjects are assigned the TT genotype using the multiple-imputation approach that incorporates information on the trait, while they are assigned the most frequent CC genotype using a single-imputation. An additional $n = 6$ subjects are assigned the TC genotype using multiple imputation and the CC genotype using the single imputation approach. In this example, knowledge about the trait informs the genotype assignment. We again note that the results may vary upon repeated applications of this procedure due to the random sampling involved in fitting random forests.                                                                        □

*Missing haplotype data*

Another form of missing data that arises in the context of genetic association studies is a result of the unobservable nature of allelic phase. A discussion of this data-analytic challenge inherent in population-based investigations, the importance of considering haplotypic phase, and several appropriate analytic tools are described in detail in Chapter 5. Here we describe one approach, termed multiple imputation and random forests (MIRF), that draws on the approaches of Excoffier and Slatkin (1995) and Breiman (2001), as well as the established statistical theory on multiple imputation as described in Little and Rubin (2002). Additional details on MIRF can be found in Nonyane and Foulkes (2007). Notably, unlike the multiple-imputation approach described above for missing genotype data, the MIRF approach does not incorporate information about the trait in reconstructing the missing data.

A step-by-step summary of the MIRF approach is given in ALGORITHM 7.3 below. The first step of this algorithm involves estimating individual-level haplotype probabilities based on the observed genotype data. That is, for each pair of haplotypes that is consistent with the observed genotype, we assign a probability that this pair is indeed the true diplotype. All other haplotype pairs have an estimated probability of zero since they are not consistent with the observed data. Here we apply the EM approach of Excoffier and Slatkin (1995) and described in detail in Section 5.1.1, though other approaches are tenable.

The second step is to generate a completed dataset based on these estimated posterior probabilities. To do this, we randomly sample a haplotype pair according to the estimated probabilities derived in step 1 of the algorithm. For example, suppose the observed genotype across two SNPs within a single gene for a given individual is $Aa$ and $Bb$. As discussed in Section 2.3.2, the two possible underlying haplotype pairs for this individual are given by $H_1 = (AB, ab)$ and $H_2 = (Ab, aB)$. Suppose the corresponding posterior probabilities are estimated to be $\widehat{p}_1 = 0.60$ and $\widehat{p}_2 = 0.40$, respectively. At this stage, we complete our data by sampling one of the two possible diplotypes with probabilities $\widehat{p}_1$ and $\widehat{p}_2$. By repeating this for each individual within our sample, we arrive at a completed dataset.

The third step involves fitting a random forest using the imputed data according to ALGORITHM 7.1 above. The overall importance scores are recorded for each of the predictor variables. We repeat steps 2 and 3 $M$ times to arrive at importance scores for each predictor variable across multiple imputed datasets. The repetition allows us to capture the variability resulting from sampling the haplotype pairs. Finally, the importance scores are then combined in a manner that takes into account the variance within and between imputations, as described in Section 5.4 of Little and Rubin (2002). Specifically, we let

$$\bar{\theta}_j = \frac{1}{M} \sum_{m=1}^{M} \widehat{\theta}_j^m \qquad (7.7)$$

be the average importance score across the $M$ imputations and define

$$T_j = \left(\bar{\theta}_j\right) V_j^{-1/2} \tag{7.8}$$

where

$$V_j = \bar{W}_j + \frac{M+1}{M} B_j \tag{7.9}$$

is a function of the variances within $(\bar{W}_j)$ and between $(B_j)$ imputations, given by

$$\bar{W}_j = \frac{1}{M} \sum_{m=1}^{M} (s_j^m)^2 \tag{7.10}$$

and

$$B_j = \frac{1}{M-1} \sum_{m=1}^{M} \left(\widehat{\theta}_j^m - \bar{\theta}_j\right)^2 \tag{7.11}$$

---

ALGORITHM 7.3: MULTIPLE IMPUTATION APPROACH TO RFS WITH MISS-
ING HAPLOTYPE DATA: We begin by initializing $m = 1$.

1. Estimate posterior diplotype probabilities for each subject and for each gene using the EM approach of Excoffier and Slatkin (1995), as described in detail in Section 5.1.1. Denote these individual-level probabilities by the vector $\mathbf{r}_{ik}$, where $i$ indicates individual, $k$ indicates gene and the elements of this vector correspond to the posterior probabilities of the diplotypes consistent with the individual's observed genotype.

2. For each individual $i$ and gene $k$, sample a single diplotype with probabilities $\mathbf{r}_{ik}$ until a complete dataset is obtained.

3. Fit a random forest according to ALGORITHM 7.1 using the dataset imputed in step (2), and record importance scores $\widehat{\theta}_j^m$ and corresponding standard errors $s_j^m$ for each predictor variable $\mathbf{x}_j$, $j = 1, \ldots, p$.

4. Repeat steps (2) and (3) $M$ times, incrementing $m$ each time, to arrive at $\widehat{\theta}_j^1, \ldots, \widehat{\theta}_j^M$ and $s_j^1, \ldots, s_j^M$.

5. Combine importance scores across multiple imputed datasets.

---

This approach is demonstrated in the following example using the `mirf()` function of the `mirf` package in R. Again it is important to reconstruct haplotypes within racial and ethnic groups since the approach of Excoffier and Slatkin (1995) assumes HWE. We can do this by first stratifying our sample prior to applying this approach or by specifying the appropriate factor variable in the `mirf()` function. It is also important to specify the genes to which the SNPs correspond since haplotypes are defined within genes, as illustrated in the example below.

*Example 7.4 (MIRF).* Suppose we are interested in determining which haplotypes within the `actn3` and `resistin` genes are most associated with change in non-dominant arm muscle strength before and after muscle training using the FAMuSS data. We begin by installing and uploading the `mirf()` package and attaching the FAMuSS data:

```
> install.packages("mirf")
> library(mirf)
> attach(fms)
```

We then specify the SNPs to be used in this analysis and subset the corresponding columns of the fms dataframe:

```
> genoSNPnames <- c("actn3_r577x","actn3_rs540874","actn3_rs1815739",
+        "actn3_1671064","resistin_c30t","resistin_c398t",
+        "resistin_g540a","resistin_c980g","resistin_c180g",
+        "resistin_a537c")
> FMSgeno <- fms[,is.element(names(fms),
+        genoSNPnames)][!is.na(NDRM.CH),]
```

The `sepGeno` function converts this genotype matrix into an object that includes two columns for each SNP, the input for `mirf()` as well as the `haplo.em()` function within the `haplo.stats` package.

```
> Geno <- sepGeno(FMSgeno)
```

Finally, we specify the trait and apply `mirf()` to the data with no missing values for the trait:

```
> Trait <- NDRM.CH[!is.na(NDRM.CH)]
> mirf(geno=Geno$geno, y=Trait, gene.column=c(8,12),
+        SNPnames=genoSNPnames, M=10)

[1] "iteration=1"
[1] "iteration=2"
[1] "iteration=3"
[1] "iteration=4"
[1] "iteration=5"
[1] "iteration=6"
[1] "iteration=7"
[1] "iteration=8"
[1] "iteration=9"
[1] "iteration=10"
    sourceGene haplotype importanceScore haplo.freq
1      actn3     CACA          -0.041       0.002
```

| 2  | actn3    | CACG   | -0.741 | 0.007 |
|----|----------|--------|--------|-------|
| 3  | actn3    | CATA   | 0.408  | 0.004 |
| 4  | actn3    | CATG   | -0.549 | 0.014 |
| 5  | actn3    | CGCA   | 1.557  | 0.48  |
| 6  | actn3    | CGCG   | -0.031 | 0.011 |
| 7  | actn3    | TATA   | 0.479  | 0.004 |
| 8  | actn3    | TATG   | 1.580  | 0.395 |
| 9  | actn3    | TGCA   | 1.070  | 0.076 |
| 10 | actn3    | TGCG   | NaN    | 0     |
| 11 | actn3    | TGTA   | NaN    | 0.001 |
| 12 | actn3    | TGTG   | -1.825 | 0.007 |
| 13 | resistin | CCACCA | -0.692 | 0.033 |
| 14 | resistin | CCACCC | NaN    | 0     |
| 15 | resistin | CCACGA | -0.220 | 0.073 |
| 16 | resistin | CCACGC | 0.374  | 0.007 |
| 17 | resistin | CCAGCA | -0.785 | 0.006 |
| 18 | resistin | CCAGGA | 0.277  | 0.001 |
| 19 | resistin | CCAGGC | 0.649  | 0.001 |
| 20 | resistin | CCGCCA | 0.703  | 0.25  |
| 21 | resistin | CCGCGA | -0.081 | 0.011 |
| 22 | resistin | CCGCGC | 0.084  | 0.037 |
| 23 | resistin | CCGGCA | 2.176  | 0.344 |
| 24 | resistin | CCGGGA | 0.025  | 0.009 |
| 25 | resistin | CCGGGC | -0.107 | 0.004 |
| 26 | resistin | CTACCA | 0.311  | 0.003 |
| 27 | resistin | CTACGA | 1.320  | 0.159 |
| 28 | resistin | CTAGGA | -0.595 | 0.017 |
| 29 | resistin | CTGCCA | -0.003 | 0.006 |
| 30 | resistin | CTGCGA | 0.428  | 0.012 |
| 31 | resistin | CTGCGC | -0.162 | 0.002 |
| 32 | resistin | CTGGCA | -0.399 | 0.008 |
| 33 | resistin | CTGGCC | NaN    | 0     |
| 34 | resistin | CTGGGA | 0.355  | 0.002 |
| 35 | resistin | CTGGGC | NaN    | 0     |
| 36 | resistin | TCACCA | 0.116  | 0.001 |
| 37 | resistin | TCGCCA | 0.579  | 0.011 |
| 38 | resistin | TTAGGA | -0.270 | 0.001 |
| 39 | resistin | TTGCCA | -0.204 | 0.001 |

Note that, in the mirf() function call, we specify gene.column=c(8,12) to indicate that the first 8 columns (4 SNPs) of our geno matrix correspond to one gene and the remaining 12 columns (6 SNPs) correspond to another gene. We also indicate that we want to perform $M = 10$ imputations. The resulting output includes the gene name, corresponding haplotype within that gene, the variable importance score as defined by the mean decrease in accuracy for randomForest(), and the estimated haplotype frequencies. Based on the output above, we see that within the resistin gene the most important haplotype is CCGGCA, with an importance score of 2.176 and an estimated frequency of

0.344 in our population. Notably, this approach assumes HWE, which may not be appropriate. Specification of HWE groups is left as an exercise for the reader.                                                                          □

### 7.1.3 Covariates

Finally, there are several different approaches for handling covariates in the RF setting, as described in Section 6.1.3 in the context of fitting a regression tree. These include (1) ignoring the covariates in the analysis, (2) including the covariates as potential predictors, (3) stratifying the analysis based on levels of the covariates and (4) residualizing the data based on prior model fitting with the covariates as predictors. Each of these methods is reasonable, and the best choice will inevitably depend on the specific scientific questions under consideration as well as the assumed underlying model of association. Importantly, recent research has illustrated the sensitivity of importance scores to the scale of the input variables and the need for conditional inference in this setting (Strobl et al., 2007). Specifically, if both categorical and continuous predictor variables are used as inputs in an RF, the results may be biased. The reader is referred to the `cforest()` function within the `party` package in R for appropriate application of RFs in this setting. For additional discussion, see Nonyane and Foulkes (2008).

## 7.2 Logic regression

Logic regression is another machine learning algorithm that is well suited to the analysis of data arising from genotype–trait association studies. General background on this approach can be found in Ruczinski et al. (2003) and Ruczinski et al. (2004), with specific applications to SNP data described in Kooperberg et al. (2001). A further extension for exploratory analysis that involves Markov chain Monte Carlo model selection is described in Kooperberg and Ruczinski (2005). Further extensions of this method, described in Schwender and Ickstadt (2008a) and Schwender and Ickstadt (2008b), provide us with the tools for measuring variable importance in a manner similar to the random forest setting. We begin by describing the logic tree framework and then discuss variable-importance measures.

Logic regression involves searching for models that consist of sums of Boolean expressions. That is, models of the form

$$g(E[\mathbf{y}]) = \beta_0 + \sum_{j=1}^{t} \beta_j L_j \qquad (7.12)$$

are considered, where $L_j$ is a Boolean combination of binary predictor variables. A Boolean expression takes on the values 0 and 1 and is a logical

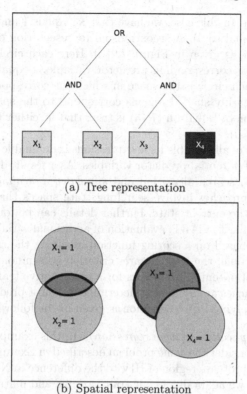

(a) Tree representation

(b) Spatial representation

**Fig. 7.2.** Example boolean statement in logic regression

function of multiple binary predictor variables. That is, it is a function of indicators of the predictor variables that involves a series of *"and"*, *"or"* and *"not"* (also referred to as *"complement"*) statements.

Consider for example the binary genotype variables $\mathbf{x}_1, \ldots, \mathbf{x}_p$, each representing an indicator for the presence of one variant allele at the corresponding sites $1, \ldots, p$. One example of a Boolean expression is given by

$$(\mathbf{x}_1 \wedge \mathbf{x}_2) \vee (\mathbf{x}_3 \wedge \mathbf{x}_4^c) \tag{7.13}$$

where $\wedge$ indicates *and*, $\vee$ indicates *or* and $c$ denotes *not*. This expression is read *"both $\mathbf{x}_1$ and $\mathbf{x}_2$ are equal to one or $\mathbf{x}_3$ is equal to one and $\mathbf{x}_4$ is equal to zero"*. A visual representation of this expression is given in Figure 7.2(a), from which we see that the expression of Equation (7.13) can be thought of as a decision tree. Above the first split, we see the **OR** statement since we can either go to the left or the right daughter node. If we follow this tree to the left, we see an **AND** statement with the two unshaded boxes for $\mathbf{x}_1$ and $\mathbf{x}_2$, indicating that both of these must equal 1. The right node of the tree also leads to an **AND** statement, with an unshaded box for $\mathbf{x}_3$ and a shaded box for $x_4$. In this case, we have that $x_3$ equals 1 and $x_4$ is not equal to 1 (i.e., $x_4$ is equal to 0). A second visual representation of the expression in Equation (7.13) is given in Figure 7.2(b). Here each circle represents the space in which the corresponding predictor variable is equal to 1, while the area outside of each circle is the space in which the corresponding variable is equal to 0. The darkly shaded regions correspond to the space in which the Boolean expression of Equation (7.13) is true; that is, either ($x_1 = x_2 = 1$) or ($x_3 = 1$ and $x_4 = 0$).

Consideration of all possible logic expressions is untenable in the context of a large number of potential predictor variables. As a result, different schemes, including a *greedy* search algorithm and *simulated annealing*, have been described. Both approaches involve searching state spaces that involve simple transitions from the current state. Further details can be found in Ruczinski *et al.* (2003). Similar to CART, evaluation of additional variables for inclusion in a logic tree is based on a scoring function involving the trait under investigation. For example, the least squares criterion is commonly used to assess the importance of potential predictors for a quantitative trait. Finally, cross-validation approaches are applied to determine the most predictive model. An illustration of applying logic regression is given in the following example.

*Example 7.5 (Application of logic regression).*  In this example, we revisit the Virco data and consider the same problem described in Example 7.1 of relating mutations in the protease region of HIV to the difference in NFV and IDV fold resistance. First, we again define our trait variable and matrix of genotypes:

```
> attach(virco)
> Trait <- NFV.Fold - IDV.Fold
> VircoGeno <- data.frame(virco[,substr(names(virco),1,1)=="P"]!="-"])
> Trait.c <- Trait[!is.na(Trait)]
> VircoGeno.c <- VircoGeno[!is.na(Trait),]
```

We then install and upload the `LogicReg` package in R:

```
> install.packages("LogicReg")
> library(LogicReg)
```

We then apply logic regression using the `logreg()` function. Here we first specify `select=1` to fit a single tree model. A plot of the resulting decision tree is obtained using the `plot()` function with a logic regression object as input. The result is illustrated in Figure 7.3 and generated using the R code

```
> VircoLogicReg <- logreg(resp=Trait.c, bin=VircoGeno.c, select=1)
> plot(VircoLogicReg)
> VircoLogicReg
```

```
score 74.654
 -261 * ((((not P20) or P36) and ((not P94) and (not P8))) or
                (((not P10) or P32) or ((not P93) or P2)))
```

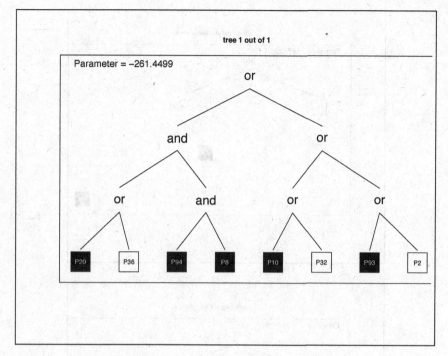

**Fig. 7.3.** Single logic regression tree from Example 7.5

The result is written formally as

$$-261.45 \times \{ [(P20^c \vee P36) \wedge (P94^c \wedge P8^c)] \\ \vee [(P10^c \vee P32) \vee (P93^c \vee P2)]\} \qquad (7.14)$$

Alternatively, we can specify select=2 to generate multiple trees within the model. For example, in the code below, we specify ntrees=2, indicating that we want a model represented by a sum of two logic trees:

```
> VircoLogicRegMult <- logreg(resp=Trait.c, bin=VircoGeno.c, select=2,
+       ntrees=2, nleaves=8)
> plot(VircoLogicRegMult)
> VircoLogicRegMult

2 trees with 8 leaves: score is 72.769
    + 188 * (P73 and ((not P54) and (P93 and (not P72))))
    - 196 * (P36 or (((not P20) or (not P93)) or P2))
```

This yields a model of the form given in Equation (7.12), where $t = 2$. Formally, this expression is given by

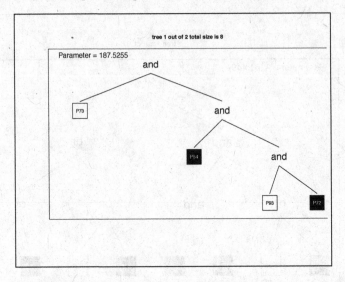

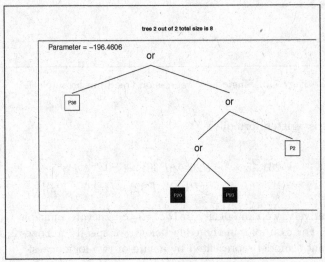

**Fig. 7.4.** Sum of logic regression trees from Example 7.5

$$187.53 \times \{[P73 \wedge (P54^c \wedge (P93 \wedge P72^c))]\}$$
$$- 196.46 \times \{P36 \vee [(P20^c \vee P93^c) \vee P2]\} \tag{7.15}$$

A visual representation of this expression is given in Figure 7.4.          □

In the example provided above, a simulated annealing algorithm is applied to generate the logic tree(s), as described in the original formulation of logic regression (Ruczinski *et al.*, 2003). One alternative is the integration of a Markov chain Monte Carlo (MCMC) algorithm in this setting, termed *Monte*

*Carlo logic regression* and proposed by Kooperberg and Ruczinski (2005). This extension was developed as an exploratory tool, with the goal of identifying multiple potential models that together offer support of an association. Coverage of the reversible jump MCMC algorithm employed is beyond the scope of this text; however, we note that application of this approach yields output that is quite different from what we saw in Example 7.5. Specifically, similar to the random forest setting, there is a shift from identifying the best model to identifying the importance of each variable or combination of variables. In this setting, importance is measured by the proportion of times a SNP is included in a selected model, among many such models. This is illustrated in the following example.

*Example 7.6 (Monte Carlo logic regression).* In this example, we return to the Virco data and apply Monte Carlo logic regression to determine which SNPs in the protease region of the viral genome are potentially associated with Saquinavir (SQV) fold resistance. We begin by uploading the `LogicReg` package and the Virco data:

```
> library(LogicReg)
> attach(virco)
```

We then define our trait as `SQV.Fold`, define the genotype matrix and subset the complete data:

```
> Trait <- SQV.Fold
> VircoGeno <- data.frame(virco[,substr(names(virco),1,1)=="P"]!="-"])
> Trait.c <- Trait[!is.na(Trait)]
> VircoGeno.c <- VircoGeno[!is.na(Trait),]
```

Finally, we apply the `logreg()` function in the same manner as in Example 7.5, with the additional option `select=7` selected to indicate that we want to employ the MCMC algorithm:

```
> VircoLogicRegMCMC <- logreg(resp=Trait.c, bin=VircoGeno.c, select=7)
```

The attribute `single` of the resulting object is a vector of length equal to the number of SNPs input and elements equal to the number of selected models in which the corresponding variable is included. A plot of the sorted values is illustrated in Figure 7.5 and generated as follows:

```
> plot(sort(VircoLogicRegMCMC$single), xlab="Sorted SNPs",
+        ylab="Number of selected models")
```

To identify the variables that stand out in this figure, we print the names of the ordered variables (results will vary):

```
> names(VircoGeno)[order(VircoLogicRegMCMC$single)]
```

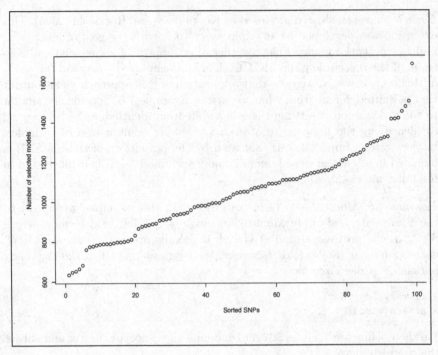

**Fig. 7.5.** Monte Carlo logic regression results from Example 7.6

```
[1]   "P25"  "P40"  "P87"  "P81"  "P56"  "P42"  "P26"  "P17"  "P59"  "P9"   "P41"
[12]  "P27"  "P94"  "P44"  "P68"  "P75"  "P38"  "P49"  "P31"  "P8"   "P29"  "P65"
[23]  "P11"  "P51"  "P96"  "P28"  "P78"  "P80"  "P37"  "P66"  "P14"  "P86"  "P30"
[34]  "P6"   "P83"  "P43"  "P2"   "P19"  "P7"   "P39"  "P21"  "P52"  "P58"  "P85"
[45]  "P16"  "P22"  "P4"   "P57"  "P89"  "P64"  "P15"  "P18"  "P74"  "P63"  "P5"
[56]  "P34"  "P69"  "P23"  "P47"  "P82"  "P70"  "P61"  "P36"  "P98"  "P97"  "P50"
[67]  "P92"  "P99"  "P13"  "P88"  "P91"  "P35"  "P77"  "P45"  "P95"  "P33"  "P62"
[78]  "P60"  "P20"  "P12"  "P55"  "P67"  "P46"  "P53"  "P76"  "P24"  "P32"  "P72"
[89]  "P71"  "P93"  "P1"   "P79"  "P3"   "P90"  "P48"  "P73"  "P54"  "P10"  "P84"
```

Here we see that the top seven variables that, based on Figure 7.5, appear to have an importance score that is notably higher than the remaining variables are mutations at sites 3, 10, 48, 54, 73, 84 and 90. Interestingly, if we compare this with the PI Resistance Notes of the HIV Drug Resistance Database (http://hivdb.stanford.edu/cgi-bin/PIResiNote.cgi), we see reasonable concordance with previous findings. Specifically, the Resistance Notes identify mutations at sites 48, 54, 73, 84 and 90 as associated with reduced *in vitro* susceptibility or *in vivo* virological response to SQV.

This implementation of Monte Carlo logic regression also provides us with information on the pairs and triplets of mutations that tend to occur in the same selected model. For example, the `double` attribute of our logic tree holds a square matrix with lower diagonal elements equal to the number of models selected that include both the corresponding row and column variables. Similarly, the `triple` attribute provides information on the number of selected models that contain each triplet of SNPs.                                   □

In a recent manuscript, Schwender and Ickstadt (2008a) propose an alternative bootstrap algorithm, termed *logicFS*, for generating a logic regression tree that yields a measure of variable importance for combinations of SNPs when the trait is dichotomous. This algorithm applies the original simulated annealing algorithm of Ruczinski *et al.* (2003) on bootstrap samples of the original data. Variable importance is defined based on how well the result tree(s) predict class membership for the out-of-bag (OOB) sample, defined as those individuals not selected in the current bootstrap. This measure is similar to the variable importance described for random forests in Section 7.1.1. Application of this approach is straightforward using the `logicFS` package within Bioconductor.

## 7.3 Multivariate adaptive regression splines

Multivariate adaptive regression splines (MARS) is an alternative machine learning approach that is closely related to CART. A complete discussion of model fitting and pruning is provided in Friedman (1991) and Hastie *et al.* (2001). Here we present the general approach and specifically illustrate the overlap between MARS and CART for the analysis of genotype–trait association. In the context of categorical predictors such as SNPs, we see that MARS is more conducive than CART to discovery of statistical interaction, though the two methods are very similar under most other models of association. For simplicity of presentation, we focus on the setting in which the genotype variables are indicators for the presence of at least one variant allele. This corresponds to a dominant genetic model, as described in Section 2.3.4, and may be a reasonable assumption with prior knowledge of functionality. Notably, if the genotypes are three-level factor variables, they can be equivalently represented by two binary variables and the approach described below is easily extended.

Throughout this section, we focus on a quantitative trait, though application of MARS to the case–control setting is straightforward. Using the same notation as we have throughout this text, suppose we are interested in characterizing the association between the set of genotype variables given by $\mathbf{x}_1, \ldots, \mathbf{x}_p$ and a trait, given by $\mathbf{y}$. Each of these are $n \times 1$ vectors, where $n$ is the number of individuals in our sample. In our setting, MARS begins by considering the model given by

$$\mathbf{Y} = \beta_0 + \beta_1(\mathbf{x}_j - t)_+ + \beta_2(t - \mathbf{x}_j)_+ + \epsilon ; \quad t \in \{x_{ij}\} \tag{7.16}$$

for each $j = 1, \ldots, p$, where $\epsilon$ is the measurement error, $t$ is an element of the set of observed values of $\mathbf{x}_j$, given by $\{x_{ij}\}$, and $()_+$ denotes the positive component of the argument within the parentheses. The components of the set $\{(\mathbf{x}_j - t)_+, (t - \mathbf{x}_j)_+\}$ are referred to in the MARS literature as reflective basis functions. In the setting of binary predictors where the $x_{ij}$ equal 0 or 1, we have that $t \in (0, 1)$ and thus Equation (7.16) can be expressed as

$$\mathbf{Y} = \beta_0 + \beta_1 \mathbf{x}_j + \epsilon \tag{7.17}$$

To see this, note that, for $t = 0$, $(\mathbf{x}_j - t)_+ = \mathbf{x}_j$ and $(t - \mathbf{x}_j)_+ = \mathbf{0}$. Furthermore, for $t = 1$, we have $(\mathbf{x}_j - t)_+ = \mathbf{0}$ and $(t - \mathbf{x}_j)_+ = 1 - \mathbf{x}_j$.

The best predictor $\mathbf{x}_j^*$ is defined as the variable that leads to the greatest reduction in the residual sums of squares. Notably, this first step is identical to the regression tree approach described in Chapter 6. In MARS, a model set $\mathcal{M}$ is then defined as the set of functions (in our case predictors) given by $\{1, \mathbf{x}_j^*\}$. The next step is to fit models that involve products of the elements of $\mathcal{M}$ and the predictor variables. That is, we next consider models of the form

$$Y = \beta_0 + \beta_1 \mathbf{x}_j^* + \beta_2 \mathbf{x}_k + \epsilon \tag{7.18}$$

and

$$Y = \beta_0 + \beta_1 \mathbf{x}_j^* + \beta_2 \mathbf{x}_j^* \mathbf{x}_k + \epsilon \tag{7.19}$$

for $k = 1, \ldots, p$.

The difference between CART and MARS becomes apparent at this stage since CART does not consider models of the form given by Equation (7.18). That is, MARS is more conducive to modeling additive structure across predictors. Again the best predictor is chosen, this time of the form $\mathbf{x}_k$ or $\mathbf{x}_j^* \mathbf{x}_k$, and then added to the model set $\mathcal{M}$. The process is repeated recursively to build a model of both additive and interaction terms. Finally, a backward deletion procedure is applied to reduce overfitting. MARS is straightforward to implement using the `earth()` function within the R package `earth`, as demonstrated in the following example.

*Example 7.7 (An application of MARS).* In this example, we again consider the Virco data, with the goal of characterizing association between mutations in the protease region of HIV and the difference in NFV and IDV fold resistance. First, we again define our trait variable matrix of genotypes:

```
> attach(virco)
> Trait <- NFV.Fold - IDV.Fold
> VircoGeno <- data.frame(virco[,substr(names(virco),1,1)=="P"]!="-")
> Trait.c <- Trait[!is.na(Trait)]
> VircoGeno.c <- VircoGeno[!is.na(Trait),]
```

We then install and upload the `earth` package:

```
> install.packages("earth")
> library(earth)
```

Applying MARS is then straightforward using the earth() function. Here we specify `degree=2` to include both main effects and two-way interaction terms as potential predictors:

```
> VircoMARS <- earth(Trait.c~., data=VircoGeno.c, degree=2)
> summary(VircoMARS)

Call: earth(formula=Trait.c~., data=VircoGeno.c, degree=2)

                         Trait.c
(Intercept)             -1.49386
P35TRUE                 36.98821
P76TRUE                -34.95785
P1TRUE * P73TRUE       -30.79950
P10TRUE * P35TRUE       29.81243
P10TRUE * P73TRUE       65.50646
P15TRUE * P25TRUE      751.24589
P15TRUE * P35TRUE      -34.54019
P15TRUE * P54TRUE       32.95728
P15TRUE * P73TRUE      -58.53545
P20TRUE * P35TRUE       47.11367
P20TRUE * P54TRUE      -41.71048
P20TRUE * P73TRUE       77.58072
P30TRUE * P70TRUE      158.97600
P30TRUE * P77TRUE       42.81780
P35TRUE * P36TRUE      -42.06393
P35TRUE * P54TRUE      -33.73524
P35TRUE * P73TRUE       78.73042
P35TRUE * P82TRUE      -31.25249
P35TRUE * P84TRUE      -59.43351
P35TRUE * P93TRUE       23.76439
P35TRUE * P95TRUE      -60.69940
P36TRUE * P54TRUE       30.17810
P36TRUE * P73TRUE     -113.98578
P48TRUE * P54TRUE      -20.80249
P54TRUE * P72TRUE       24.06139
P54TRUE * P73TRUE      -63.96128
P54TRUE * P84TRUE       34.96787
P54TRUE * P93TRUE      -18.74152
P54TRUE * P94TRUE      207.51818
P63TRUE * P73TRUE       67.33288
P70TRUE * P73TRUE     -103.04692
P72TRUE * P73TRUE      -69.71491
P73TRUE * P74TRUE      -54.83226
P73TRUE * P76TRUE      101.72366
P73TRUE * P77TRUE      -54.40373
P73TRUE * P84TRUE      -65.68984
P73TRUE * P93TRUE       49.44217

Selected 38 of 100 terms, and 22 of 99 predictors
```

```
Importance: P15TRUE, P25TRUE, P35TRUE, P36TRUE, P73TRUE, P54TRUE,
            P94TRUE, P10TRUE, P77TRUE, P84TRUE, ...
Number of terms at each degree of interaction: 1 2 35
GCV 5155.408    RSS 4113795    GRSq 0.2200334    RSq 0.3610069
```

The output above includes the coefficients for each term in the final model. It also indicates that 38 terms were initially selected and 22 remained after pruning (including the intercept). The order of importance of the input variables is given by applying the evimp() function:

```
> evimp(VircoMARS)
```

| | col | used | nsubsets | gcv | | rss | |
|---|---|---|---|---|---|---|---|
| P15TRUE | 15 | 1 | 37 | 100.000000 | 1 | 100.000000 | 1 |
| P25TRUE | 25 | 1 | 37 | 100.000000 | 1 | 100.000000 | 1 |
| P35TRUE | 35 | 1 | 35 | 65.189479 | 1 | 76.166293 | 1 |
| P36TRUE | 36 | 1 | 35 | 65.189479 | 1 | 76.166293 | 1 |
| P73TRUE | 73 | 1 | 35 | 65.189479 | 1 | 76.166293 | 1 |
| P54TRUE | 54 | 1 | 33 | 50.930070 | 1 | 65.218336 | 1 |
| P94TRUE | 94 | 1 | 33 | 50.930070 | 1 | 65.218336 | 1 |
| P10TRUE | 10 | 1 | 33 | 50.071147 | 1 | 64.536462 | 1 |
| P77TRUE | 77 | 1 | 33 | 50.071147 | 1 | 64.536462 | 1 |
| P84TRUE | 84 | 1 | 32 | 47.785561 | 1 | 61.929011 | 1 |
| P72TRUE | 72 | 1 | 30 | 37.117366 | 1 | 53.182110 | 1 |
| P93TRUE | 93 | 1 | 29 | 33.045941 | 1 | 49.600900 | 1 |
| P20TRUE | 20 | 1 | 28 | 28.885308 | 1 | 45.995833 | 1 |
| P82TRUE | 82 | 1 | 25 | 15.890025 | 1 | 35.057174 | 1 |
| P30TRUE | 30 | 1 | 23 | 11.377226 | 1 | 30.262470 | 1 |
| P1TRUE | 1 | 1 | 21 | 9.000836 | 1 | 26.741314 | 1 |
| P70TRUE | 70 | 1 | 21 | 8.952997 | 1 | 26.704566 | 1 |
| P74TRUE | 74 | 1 | 15 | 4.663905 | 1 | 17.893339 | 1 |
| P48TRUE | 48 | 1 | 13 | 4.491108 | 1 | 16.017860 | 1 |
| P95TRUE | 95 | 1 | 13 | 4.103631 | 1 | 15.479906 | 1 |
| P85TRUE-unused | 85 | 0 | 11 | 3.526339 | 1 | 13.431009 | 1 |
| P89TRUE-unused | 89 | 0 | 10 | 2.924444 | 1 | 12.030094 | 1 |
| P63TRUE | 63 | 1 | 7 | 2.587980 | 1 | 8.452929 | 1 |
| P65TRUE-unused | 65 | 0 | 7 | 1.777517 | 1 | 8.221570 | 1 |
| P76TRUE | 76 | 1 | 5 | 2.175851 | 0 | 6.167241 | 1 |
| P45TRUE-unused | 45 | 0 | 1 | -0.031605 | 1 | 1.008404 | 1 |

Those variables included in the final model are indicated with a 1 in the column entitled used. Comparing the MARS results with the results of fitting RF in Example 7.1 and applying logic regression in Example 7.5 to the same data reveals the potential for variability in the findings of these three machine learning algorithms. While many of the same variables stand out across all three analyses (e.g., $P20$, $P36$, $P54$, $P73$ and $P94$), several other variables appear important based on one analysis but do not according to another analysis. For example, we see that $P10$ and $P93$ are ranked relatively low using RF compared with both MARS and logic regression.    □

# 7.4 Bayesian variable selection

In Chapter 5, we introduced Bayesian methods in the context of haplotype reconstruction. More broadly, Bayesian approaches offer an alternative framework for evaluating a large number of potential models of association between multiple genotypes and a trait. In this section, we introduce Bayesian variable selection as described in the seminal paper by George and McCulloch (1993). A growing body of literature exists that further extends these methods and is widely applicable to data arising from population-based genetic association studies. Here we describe fundamental concepts that offer the reader tools for further explorations.

In the study of genotype–trait associations, our goal is often to identify a model of the form

$$y_i = x_{i1}^* \beta_1 + x_{i2}^* \beta_2 + \ldots + x_{ir}^* \beta_r + \epsilon_i \qquad (7.20)$$

for $i = 1, \ldots, n$, where $(\mathbf{x}_1^*, \ldots, \mathbf{x}_r^*)$ is a subset of the potential predictor variables, $\mathbf{y}$ is a quantitative trait and we typically assume the $\epsilon_i$'s are independent and identically distributed, $N(0, \sigma^2)$. Traditional statistical methods may involve fitting a series of regression equations, each involving a different subset, and comparing these models. The number of such models, given by $2^p$, is large, however, and consideration of all such models is computationally infeasible in many instances. The goal of Bayesian variable selection (BVS) is to identify a subset of promising variables that have a high probability of being associated with the trait and are thus worthy of further consideration.

We begin by writing the full model of Equation (7.20) (that is, the model with all $p$ potential predictor variables) in matrix notation as

$$\mathbf{y} | \boldsymbol{\beta}, \sigma^2 \sim MVN_n(\mathbf{X}\boldsymbol{\beta}, \sigma^2 I) \qquad (7.21)$$

where $\mathbf{y} = (y_1, \ldots, y_n)^T$, $\mathbf{X}_{n \times p} = [\mathbf{x}_1, \ldots, \mathbf{x}_p]$ and $\boldsymbol{\beta} = (\beta_1, \ldots, \beta_p)^T$. Notably, the parameters corresponding to the true underlying predictors $(\mathbf{x}_1^*, \ldots, \mathbf{x}_r^*)$ of Equation (7.20) will be non-zero while the parameters corresponding to all other $\mathbf{x}_j$'s will be identically equal to 0.

We have seen the model of Equation (7.21) in Section 2.2.3, and the reader will recognize it as the standard linear regression model. The BVS approach extends this setting by introducing a new component to this model. Namely, the parameters in the model, given by $\boldsymbol{\beta}$ and $\sigma^2$, are treated as random variables, arising from a known distribution (with potentially unknown parameters). Specifically, a normal mixture model is assumed for each $\beta_j$, given by

$$\beta_j | \gamma_j \sim (1 - \gamma_j) N(0, \tau_j^2) + \gamma_j N(0, c_j^2 \tau_j^2) \qquad (7.22)$$

where $\boldsymbol{\gamma} = (\gamma_1, \ldots, \gamma_p)$ is a latent (unobservable) vector with elements taking on the values 0 and 1. This is equivalent to assuming $\beta_j$ comes from one of two normal distributions with probabilities $Pr(\gamma_j = 1) = 1 - Pr(\gamma_j = 0) = p_j$ and $1 - p_j$. Here $\tau_j$ is assumed to be small so that $\beta_j \sim N(0, \tau_j^2)$ will be approximately 0. On the other hand, $c_j$ is assumed to be relatively large so that $\beta_j \sim N(0, c_j^2 \tau_j^2)$ is non-zero. We similarly make an assumption about the distribution of the variance parameter from our model. Specifically, we let

$$\sigma^2 | \gamma \sim IG(\nu_\gamma/2, \nu_\gamma \lambda_\gamma/2) \tag{7.23}$$

In general, interest lies in the posterior distribution of $\gamma$ given the observed data since this provides us with information about which variables are predictive of the trait under investigation. That is, our aim is to characterize

$$\pi(\gamma | \mathbf{Y}) \propto f(\mathbf{Y} | \gamma) \pi(\gamma) \tag{7.24}$$

where the right-hand side of Equation (7.24) is derived using Bayes' rule as described in Section 5.1.2. Using a Gibbs sampler, we can draw from this distribution by repeated sampling from the marginal posterior densities of $\beta$, $\sigma$ and $\gamma_j$. A step-by-step procedure is given by ALGORITHM 7.4, where we begin by letting the counter $t = 0$.

---

ALGORITHM 7.4: GIBBS SAMPLER FOR BAYESIAN VARIABLE SELECTION

1. Initialize $\beta$, $\sigma$ and $\gamma$ and denote these by $\beta^{(0)}$, $\sigma^{(0)}$ and $\gamma^{(0)}$.

2. Let $t = t + 1$ and sample:

- $\beta^{(t)} | \mathbf{y} \sim f(\beta | \mathbf{y}, \sigma^{(t-1)}, \gamma^{(t-1)})$

- $\sigma^{(t)} | \mathbf{y} \sim f(\sigma | \mathbf{y}, \beta^{(t)}, \gamma^{(t-1)})$

3. Randomly select an ordering $\gamma_{(1)}, \ldots, \gamma_{(p)}$ and sample:

- $\gamma_{(1)}^{(t)} | \mathbf{y} \sim f(\gamma_{(1)} | \mathbf{y}, \beta^{(t)}, \sigma^{(t)}, \gamma_{(2)}^{(t-1)}, \ldots, \gamma_{(p)}^{(t-1)})$

- $\gamma_{(2)}^{(t)} | \mathbf{y} \sim f(\gamma_{(2)} | \mathbf{y}, \beta^{(t)}, \sigma^{(t)}, \gamma_{(1)}^{(t)}, \gamma_{(3)}^{(t-1)}, \ldots, \gamma_{(p)}^{(t-1)})$

$\vdots$

- $\gamma_{(p)}^{(t)} | \mathbf{y} \sim f(\gamma_{(p)} | \mathbf{y}, \beta^{(t)}, \sigma^{(t)}, \gamma_{(1)}^{(t)}, \ldots, \gamma_{(p-1)}^{(t)})$

4. Repeat steps (2) and (3) $M$ times for $M$ large.

---

Specific forms of the distributions in the sampler above, as well as details on sensitivities to model specifications, are described in George and McCulloch (1993). Once a stationary distribution is achieved, the empirical distribution of $\gamma$ will approximate the posterior distribution, $\pi(\gamma | \mathbf{y})$. Therefore, the variables

$\mathbf{x}_j$ for which the corresponding $\gamma_j$ has a high frequency of equaling 1 can be selected to form a promising subset of predictive variables.

A few approaches have been described recently for genome-wide association studies that draw on BVS methods. For example, Lunn *et al.* (2006) describe a Bayesian toolkit for genome-wide association (GWA) studies, with associated WinBUGS software, and Schumacher and Kraft (2007) describe a Bayesian latent class analysis for GWA. The Bayesian additive regression tree (BART) approach is a further extension described by Chipman *et al.* (2008) that can be implemented in R using the `BayesTree` package. Hoggart *et al.* (2008) propose a "Bayesian inspired" approach for simultaneous consideration of a large set of SNPs potentially associated with a binary trait that applies a stochastic search algorithm to address the computational challenge of whole and partial genome-wide association studies.

## 7.5 Further readings

A broad array of literature exists on data-mining methods that are beyond the scope of the present text. In this chapter, we described a handful of well-defined algorithms for exploring high-dimensional data. These methods were chosen for presentation based on their strong theoretical foundations and straightforward applications to data arising from genetic association studies, as illustrated in the examples presented throughout. As indicated, this coverage is not intended to be comprehensive; instead, an introduction to these methods will hopefully entice the motivated reader to explore these topics further. Cogent summaries of these methods for the more advanced reader are provided in Hastie *et al.* (2001) and Chapter 16 of Gentleman *et al.* (2005).

Additional topics that may serve as useful tools in the analysis of genotype–trait association in population-based investigations include Bootstrap AGGre-gatING (Breiman, 1996); boosting (Schapire, 1990; Freund, 1990); AdaBoost (Freund and Schapire, 1997); bump hunting (Friedman and Fisher, 1999); support vector machines (Christianini and Shawe-Taylor, 2000); ridge regression; the lasso (Tibshirani, 1996) and elastic net (Zou and Hastie, 2005), among others. Summaries of the application of several data-mining approaches to genetic association data are given in Cupples *et al.* (2005) and Costello *et al.* (2003). While these approaches are well described for knowledge discovery, further investigation of their applicability to data arising from population-based investigations is needed. Specifically, careful consideration of the potential interplay among variables, as described in Section 2.1.2, and how these machine learning algorithms handle such inputs, as well as the reproducibility of any single analysis is essential to arrive at biologically and clinically interpretable findings.

# Problems

**7.1.** Fit a random forest to the Virco data to determine which protease mutations are most highly associated with SQV fold resistance as measured by SQV.fold.

**7.2.** Apply logic regression to the Virco data to characterize, with a logic structure, the association between protease mutations and SQV fold resistance as measured by SQV.fold.

**7.3.** Apply Monte Carlo logic regression to the FAMuSS data to characterize the importance of SNPs in the actn3 and resistin genes in predicting change in non-dominant arm muscle strength as measured by NDRM.CH.

**7.4.** Apply a multivariable adaptive regression spline to address the question of Problem 7.3. Compare and contrast your findings.

# Appendix

## R Basics

The purpose of this appendix is to introduce the fundamental concepts of programming in R that will provide sufficient fluency for the reader to understand the examples presented throughout this text. Specifically, we offer an overview of importing and manipulating data, installing and using R packages, and writing and applying basic functions. The novice reader is encouraged to refer to a number of comprehensive texts for a more thorough introduction to working and programming in R. See for example Gentleman (2008), Venables and Smith (2008), Spector (2008) and Dalgaard (2002).

## A.1 Getting started

R is a free, open-source statistical computing language and environment available for Windows, Mac OS X, Linux and Unix and licensed under the terms of the GNU (GNU is Not Unix) General Public License (Free Software Foundation, 2007). R is available on the Comprehensive R Archive Network (CRAN) website (http://cran.r-project.org/), and installation is straightforward. We recommend downloading the precompiled binary distribution that is suitable to your operating system and then following the corresponding instructions.

Java GUI for R (JGR—pronounced "jaguar") is a graphical user interface (GUI) that is also freely available for download. It is distributed in binary form for Windows and Mac OS X 10.4.4 (and above) users at http://jgr.markushelbig.org/Download.html and can be compiled from source code for other platforms. JGR is not necessary to get started with using R but may be preferable to users more familiar with Windows-based applications. JGR has the primary advantage over other available GUIs of being platform independent. More information on the features and unique attributes of JGR can be found in Helbig et al. (2005).

213

*Command line*

Upon opening R (or JGR), you will see an R console that lists the version number and other pertinent information, including the current working directory, as shown below:

```
R version 2.7.1 (2008-06-23)
Copyright (C) 2008 The R Foundation for Statistical Computing
ISBN 3-900051-07-0

R is free software and comes with ABSOLUTELY NO WARRANTY.
You are welcome to redistribute it under certain conditions.
Type 'license()' or 'licence()' for distribution details.

  Natural language support but running in an English locale

R is a collaborative project with many contributors.
Type 'contributors()' for more information and
'citation()' on how to cite R or R packages in publications.

Type 'demo()' for some demos, 'help()' for on-line help, or
'help.start()' for an HTML browser interface to help.
Type 'q()' to quit R.

[Workspace restored from /Users/foulkes/.RData]

>
```

Commands can be typed directly at the prompt, indicated by the > symbol, or saved in an ascii file using any text editor. Since R is an interpreter language, each line of code is processed as it is entered. This feature is similar to alternative statistical programming languages such as STATA (http://www.stata.com/) and distinguishes R from compiler languages such as C/C++ and Fortran. Notably, this allows us to transition seamlessly in R between sourcing text files and typing commands directly at the prompt.

As a user, you have the choice of writing commands directly at the prompt or storing them in text files that can be sourced. In general, we strongly encourage the reader to maintain good programming practices by writing and saving code in organized, well-documented files. This is essential for generating reproducible findings, an ethical imperative in medical research studies involving human subjects. Consider, for example, the canonical "Hello World" program that simply prints the words "hello world". One approach in R is to simply type the command

```
> print("hello world")
```

directly at the prompt, which leads to the following output:

```
[1] "hello world"
```

Alternatively, we can create a text file entitled `hello_world.r`, within which each line represents a command to be executed. For example, this program might contain the text `print("hello world")`. Any lines beginning with the # symbol within this file are treated as comments and ignored in the R execution. Suppose this file is saved in the directory ~/Projects/ASG/. Using the `source()` function, we call up this file at the R prompt by submitting

```
> source("~/Projects/ASG/hello_world.r")
```

which results in the same output as before:

```
[1] "hello world"
```

It is also straightforward to run this program at a Unix or DOS prompt using the following command line within the directory in which `hello_world.r` is stored:

```
$ R CMD BATCH hello_world.r hello_world.out
```

This results in the creation of a new file entitled `hello_world.out` that contains the R output.

*Working directory*

At the beginning of an R session, it is important to identify and indicate the directory in which you want your data to be stored. This will provide for easy access to these files during subsequent sessions and facilitate working on multiple projects. It is helpful to use different directories for each project since creating new objects of the same name will overwrite existing objects. Once a directory is specified, the data will be stored in a file entitled .RData. Suppose for example that we want our working directory to be ~/Projects/ASG/Examples/. First we check the current working directory using the `getwd()` function as follows:

```
> getwd()
```

```
[1] "/Users/foulkes"
```

To change the current working directory, we apply the `setwd()` function:

```
> setwd("~/Projects/ASG/Examples/")
```

Any new objects (variables, dataframes, etc.) that you create will be saved in the workspace entitled .RData within the current working directory. The next time you begin an R session, this workspace can be loaded using the `load()` function as follows:

```
> load("~/Projects/ASG/Examples/.RData")
```

## A.2 Types of data objects

There are several different types of objects that you may be creating or manipulating during an R session, including vectors, factors, matrices, lists and dataframes. Here we briefly describe the characteristics of each with simple demonstrations of how they can be created and viewed. In Section A.3, we describe approaches to reading in data from an existing file.

*Assignment operator*

A fundamental component of programming in R is the *assignment* operator. Suppose, for example, that we want to create a new variable named `NewVar` that is equal to 1. We do this using the `<-` symbol, as demonstrated in the following code:

```
> NewVar <- 1
```

Note that the variable name is on the left-hand side of the assignment arrow and the value that we want to assign to this variable name is on the right-hand side. This ordering is not necessary in R but is conventional and recommended. Alternatively, the = symbol can be used for assignment; however, this is highly discouraged. The resulting object can be printed using the `print()` function as

```
> print(NewVar)

[1] 1
```

or more simply

```
> NewVar

[1] 1
```

*Object class*

Every object has an associated *class* that defines how functions are applied to it. Two simple classes are `numeric` and `character`. For example, the variable `NewVar` that we just defined is a `numeric` variable, which we can see by applying the `class()` function:

```
> class(NewVar)

[1] "numeric"
```

A `character` variable is created using quotation marks, as shown below:

```
> CharVar <- "Gene1"
> class(CharVar)

[1] "character"
```

Other object classes include `factors`, `matrices` and `data.frames`, each of which is described below. Throughout this text, we create objects of many different classes, ranging from linear models generated using the `lm()` function in Chapter 2 to trees generated using the `rpart()` function in Chapter 6. It is important to keep in mind that the result of applying standard R functions, such as `print()` and `plot()`, will depend on the class associated with the argument to these functions. For example, in Example 6.2, we use the `plot()` function to plot the results of a classification tree, shown in Figure 6.2. We use the same `plot()` function in Example 7.5 to plot the results of applying logic regression, which yields the very different illustration in Figure 7.3.

*Vectors*

A *vector* is defined simply as a string of objects and can be generated using the `c()` function in R. For example, we can create the vector $\mathbf{y} = (1, 2, 3)$ using the command

```
> y <- c(1,2,3)
> y
```

```
[1] 1 2 3
```

The dimension of the resulting vector is unspecified, taking on the value $1 \times 3$ or $3 \times 1$ depending on the specific application, as demonstrated below. The `c()` function can take as its input both numeric and character objects, though if both are passed to it simultaneously, they will be coerced to be of the same type. For example, the command

```
> c(1,2,3,"gene")
```

yields

```
[1] "1"     "2"     "3"     "gene"
```

In the context of genetic association studies, the trait under investigation is often represented as a vector. For example, suppose we are interested in determining the genetic contributors to an abnormal lipid profile. In this case, the trait might be total cholesterol level and is represented by the $n \times 1$ vector $\mathbf{y}$, where $n$ is the number of individuals in our sample. Each element of this vector is the cholesterol level for the corresponding individual in our sample.

*Factors*

A factor object is a type of vector with an associated attribute that describes the levels of the variable and can be created using the R function `factor()`. For example, suppose we have a vector of genotypes given by $\mathbf{x}_1^T =$ (AA, AT, TT, TT, AT, AT, AA, TT), with each element corresponding to the genotype of one individual in our sample of $n = 8$ observations. A detailed description of genotype data is given in Section 1.3. We can generate a factor object corresponding to these data using the following code:

218    Appendix R Basics

```
> x1 <- factor(c("AA", "AT", "TT", "TT", "AT", "AT", "AA", "TT"))
> x1

[1] AA AT TT TT AT AT AA TT
Levels: AA AT TT
```

We see from the output that there are three levels to this variable, given by
AA, AT and TT. Notably, use of factor objects as predictor variables in a regression setting must proceed with caution since interpretation of associated
coefficients is not always straightforward. For example, we typically code gender as a numeric vector with elements equal to 1 for males and 0 for females.
If gender is converted to a factor variable, then it is treated by the lm() function in R as taking on the values 0.5 and $-0.5$ in the regression equation.
Calculating predicted responses for males and females based on the output of
this function requires consideration of this alternative coding.

*Matrices*

A matrix is a two-dimensional array of objects and can be generated in R
in multiple ways. Consider for example the matrix of genotype data given by
$\mathbf{X} = [\mathbf{x}_1, \mathbf{x}_2]$, where $\mathbf{x}_1$ is defined above and $\mathbf{x}_2^T =$(GG, GG, GC, CC, GG, CC,
CC, GC). Each row of $\mathbf{X}$ represents an individual and each column represents
a site on the genome. This matrix can be created in R using the matrix()
function as follows:

```
> X <- matrix(c("AA", "AT", "TT", "TT", "AT", "AT", "AA", "TT",
    "GG", "GG", "GC", "CC", "GG", "CC", "CC", "GC"), nrow=8)
> X

     [,1] [,2]
[1,] "AA" "GG"
[2,] "AT" "GG"
[3,] "TT" "GC"
[4,] "TT" "CC"
[5,] "AT" "GG"
[6,] "AT" "CC"
[7,] "AA" "CC"
[8,] "TT" "GC"
```

As we see above, the input to the matrix() function is a vector containing
the elements of the matrix, beginning in the top left corner, going down the
first column and then beginning again at the top of the next column. We also
specify nrow=8 to indicate the number of rows in the matrix. Alternatively, we
can generate the vectors $\mathbf{x}_1$ and $\mathbf{x}_2$ and use the cbind() function, as follows:

```
> x1 <- c("AA", "AT", "TT", "TT", "AT", "AT", "AA", "TT")
> x2 <- c("GG", "GG", "GC", "CC", "GG", "CC", "CC", "GC")
> X <- cbind(x1, x2)
> X
```

```
       x1    x2
[1,]  "AA"  "GG"
[2,]  "AT"  "GG"
[3,]  "TT"  "GC"
[4,]  "TT"  "CC"
[5,]  "AT"  "GG"
[6,]  "AT"  "CC"
[7,]  "AA"  "CC"
[8,]  "TT"  "GC"
```

The characteristics of our resulting matrix can be printed using the `attributes()` function. For example, for our genotype matrix $\mathbf{X}$, we have

```
> attributes(X)
```

```
$dim
[1] 8 2
```

```
$dimnames
$dimnames[[1]]
NULL
```

```
$dimnames[[2]]
[1] "x1" "x2"
```

This output tells us that the dimension of $\mathbf{X}$ is $8 \times 2$ (eight rows and two columns) and the names of the two columns are given by `x1` and `x2`, respectively. The names of the rows are not specified, and hence the `NULL` value is returned for this parameter. Each attribute can be printed separately by simply typing `attributes(X)$` followed by the name of the attribute. For example, the dimension of $\mathbf{X}$ is given by

```
> attributes(X)$dim
```

```
[1] 8 2
```

*Lists*

Another useful object type in R is a list. In fact, the `dimnames` attribute of $\mathbf{X}$ in the previous example is stored as a list. A list can contain multiple objects of different types, including vectors, matrices, and lists. For example, using the `list()` function, we can generate a list that contains the vector $\mathbf{y}$ and the matrix $\mathbf{X}$ that we just created above:

```
> list(trait=y, genotypes=X)
```

```
$trait
[1] 1 2 3
```

```
$genotypes
```

```
      X1   X2
[1,]  "AA" "GG"
[2,]  "AT" "GG"
[3,]  "TT" "GC"
[4,]  "TT" "CC"
[5,]  "AT" "GG"
[6,]  "AT" "CC"
[7,]  "AA" "CC"
[8,]  "TT" "GC"
```

*Dataframes*

Dataframes are similar to matrices, with the notable exception that they can have columns of different variable types, including numeric, character and factor variables. This characteristic makes dataframes extremely useful in the analysis of population-level data in which both continuous and categorical variables need to be stored for analysis. As a result, dataframes are used extensively throughout this text.

## A.3 Importing data

All of the examples in this text are based on data that are stored in tab- or comma-delimited ascii text files. In general, data can be exported from other programs into this format and then read into R. For example, if your data are currently in a Microsoft Excel spreadsheet, simply open this file, go to the File menu and select Save As. Under the Format tab, choose CSV (comma delimited) or Text (tab delimited). Click Save, and the resulting file can now be imported into R.

In Section 1.3.3 of this text, we illustrate how to use the read.delim() and read.csv() functions in R to import tab- and comma-delimited files, respectively. For example, consider the tab-delimited text file entitled "FMS_data.txt". We import the data into R using the following commands:

```
> fmsURL <- "http://people.umass.edu/foulkes/asg/data/FMS_data.txt"
> fms <- read.delim(file=fmsURL, header=T, sep="\t")
```

The default settings for the read.delim() function are header=T, which specifies that the first row of the text file consists of variable names, and sep="\t", which indicates the data are tab delimited. Alternatively, we can specify header=F, which assumes the first line of the data file is the first patient record. If the variables are separated by columns, we specify sep=",", and if they are separated by one or more spaces, we use sep="". Alternatively, if the data are stored in a SAS export file, the sasxport.get() function of the Hmisc package can be used to read the data into R. A discussion of how to install R packages is given below. See documentation for the Hmisc package on the CRAN website for more details on using the sasxport.get() function and linking to associated macros.

# A.4 Managing data

R is a powerful and versatile tool for data management. Here we introduce some useful commands that allow basic data manipulations. Consider first the following dataframe, consisting of identification numbers, genotypes, genders and disease statuses of five subjects:

```
> SampleDat <- data.frame(ID = c(1,2,3,4,5),
+   SNP=c("AA", "AT", "TT", "TT", "AA"),
+   Gender=c("Female", "Male", "Female", "Female", "Male"),
+   DiseaseStatus=c(1, 1, 0, 0, 0))
> SampleDat

  ID SNP Gender DiseaseStatus
1  1  AA Female             1
2  2  AT   Male             1
3  3  TT Female             0
4  4  TT Female             0
5  5  AA   Male             0
```

We can print data on each variable of the resulting dataframe by using the $ symbol as follows:

```
> SampleDat$ID

[1] 1 2 3 4 5

> SampleDat$SNP

[1] AA AT TT TT AA
Levels: AA AT TT
```

Alternatively, we can first apply the attach() function to our dataframe

```
> attach(SampleDat)
```

which allows us to use the shorthand notation

```
> ID

[1] 1 2 3 4 5

> SNP

[1] AA AT TT TT AA
Levels: AA AT TT
```

The names of all of the variables in our dataframe can be listed using the names() function:

```
> names(SampleDat)

[1] "ID"            "SNP"          "Gender"        "DiseaseStatus"
```

Now suppose we want to determine the number of individuals and the number of variables in our dataset. We can do this using the `dim()` function

```
> dim(SampleDat)
```

```
[1] 5 4
```

which tells us that we have five individuals (rows) and four variables (columns). Tabulating the number of males and females is also straightforward using the `table()` function:

```
> table(SampleDat$Gender)
```

```
Female    Male
    3       2
```

Here we see that the `table()` function results in a vector with an element for each level of the input variable. Specifically, this output tells us that there are three females and two males in the dataset. The frequency of males and females in our sample is calculated using any of the equivalent commands

```
> table(SampleDat$Gender)/5
> table(SampleDat$Gender)/dim(SampleDat)[1]
> table(SampleDat$Gender)/sum(table(SampleDat$Gender))
```

which all yield the following output:

```
Female    Male
  0.6     0.4
```

Several useful aspects of R were applied in the code above. First we note that we applied division to every element of the vector given by `table()` by simply dividing by a scalar, in this case 5. We also see that we can pull out an element of a vector using [ ] with a corresponding index. Recall that applying the `dim()` function yielded a vector with the first element equal to the number of rows and the second element equal to the number of columns. By specifying [1], we are pulling out the first element of this vector, which is again the number 5. We revisit this below. Finally, by applying the function `sum()` to our table vector, we get the sum of the elements of this vector, given in this case by $3 + 2 = 5$.

We see from the results above that the gender variable is coded here as a character variable. We may want to create a new variable for gender that is coded as a numeric variable, which we can do using the `as.numeric()` function:

```
> GenderNum <- as.numeric(SampleDat$Gender)
> GenderNum
```

```
[1] 1 2 1 1 2
```

If we want males to be coded as 1 and females to be coded as 0, we simply subtract 1 from this vector. Again, we can subtract a scalar from a vector and it will apply to all elements of the vector, as shown below:

```
> GenderNum-1

[1] 0 1 0 0 1
```

Subsetting a dataframe can be a useful tool and can be achieved similar to the example above of pulling out elements of a vector. For example, suppose we want to print out the first row of our data. We can do this by specifying

```
> SampleDat[1,]

  ID SNP Gender DiseaseStatus
1  1  AA Female             1
```

Note that the number before the comma within the brackets indicates the row number. A number after the comma indicates the column number. For example, we can print the third column using

```
> SampleDat[,3]

[1] Female Male   Female Female Male
Levels: Female Male
```

Multiple rows and columns can also be printed by replacing the scalar with a sequence of numbers. For example, we can print the second and fourth columns by using the following command:

```
> SampleDat[,c(2,4)]

  SNP DiseaseStatus
1  AA             1
2  AT             1
3  TT             0
4  TT             0
5  AA             0
```

Alternatively, we can indicate one or more rows to be printed by specifying the level of a variable. For example, suppose we want to print the records for females. This is achieved as follows:

```
> SampleDat[SampleDat$Gender=="Female",]

  ID SNP Gender DiseaseStatus
1  1  AA Female             1
3  3  TT Female             0
4  4  TT Female             0
```

Subsetting our data is especially useful if we aim to do a stratified analysis. For example, suppose we want to tabulate genotypes for those individuals with disease and those without disease. One approach is to tabulate the data for the two disease statuses separately. Using the table() function in R, we have

```
> table(SampleDat $SNP[SampleDat$DiseaseStatus==1])
```

```
AA AT TT
 1  1  0
```

```
> table(SampleDat $SNP[SampleDat$DiseaseStatus==0])
```

```
AA AT TT
 1  0  2
```

Alternatively, we can use the `tapply()` function to calculate the table means simultaneously:

```
> tapply(SampleDat$SNP, SampleDat$DiseaseStatus, table)
```

```
$'0'
```

```
AA AT TT
 1  0  2
```

```
$'1'
```

```
AA AT TT
 1  1  0
```

The result is a list with the number of elements equal to the number of levels of `DiseaseStatus` and each element representing the result of applying the `table()` function to the corresponding subset of the data.

## A.5 Installing packages

Throughout this text, we use several R packages that contain useful functions for the analysis of data arising from genetic association studies. Many of these are not included in a standard download and need to be downloaded and installed, which can be done using the `install.package()` function. For example, suppose we want to use a function within the `genetics` package. We simply type

```
> install.packages("genetics")
```

Once a prompt appears, select a CRAN mirror in a location near you to download the package and the package will be installed automatically. To access the functions within the package, we need to use the `library()` function at the start of each new R session. For example, to load the `genetics` package, we write

```
> library(genetics)
```

Bioconductor is a development project that distributes several R packages targeted for the analysis of genomic data. Additional information on the aims and scope of this project as well as its practical applications can be found in Gentleman *et al.* (2004) and Gentleman *et al.* (2005). Initially this project focused on applications in DNA microarray studies, but it has recently expanded to include useful tools for the analysis of SNP data. To install Bioconductor packages, simply type

```
> source("http://bioconductor.org/biocLite.R")
> biocLite()
```

This installation can take several minutes. Additional packages can be installed by specifying the names of the packages within the `biocLite()` function

```
> source("http://bioconductor.org/biocLite.R")
> biocLite(c("pckg1","pckg2"))
```

where `pckg1` and `pckg2` are the names of packages. Additional information about Bioconductor can be found at http://www.bioconductor.org/.

## A.6 Additional help

The CRAN website's FAQs are an excellent source of information for questions about R. In addition, you can get help on how to use existing commands using the `help()` function. For example, if you want more information on the `read.table()` function, you can enter

```
> help(read.table)
```

The associated help file includes a general description of the function and how to use it, details on the arguments to be entered in the function call and the value(s) returned by it. In addition, most documentation includes at least a simple example illustrating application of the function. Finally, if you do not know the name of a function, you can use the `help.search()` command to search the existing documentation for character strings that match your input. For example, typing

```
> help.search("variance")
```

will yield a list of functions and associated packages that include the term variance in their documentation.

# References

Abecasis, G., Cherny, S., Cookson, W., and Cardon, L. (2001). GRR: graphical representation of relationship errors. *Bioinformatics*, **17**(8), 742–743.

Affymetrix (2006). Technical data sheet. *www.affymetrix.com/support/technical/datasheets/500k_datasheet.pdf*.

Agresti, A. (2002). *Categorical Data Analysis*. John Wiley & Sons.

Ahlbom, A. and Alfredsson, L. (2005). Interaction: A word with two meanings creates confusion. *European Journal of Epidemiology*, **20**, 563–564.

Alberts, B., Bray, D., Lewis, J., Raff, M., Roberts, K., and Watson, J. (1994). *Molecular Biology of the Cell*. Garland Publishing, Taylor & Francis Group.

Balding, D. (2006). A tutorial on statistical methods for population association studies. *Nature Reviews Genetics*, **7**, 781–791.

Benjamini, Y. and Hochberg, Y. (1995). Controlling the false discovery rate: A practical and powerful approach to multiple testing. *Journal of the Royal Statistical Society, Series B: Methodological*, **57**, 289–300.

Benjamini, Y. and Yekutieli, D. (2001). The control of the false discovery rate in multiple testing under dependency. *The Annals of Statistics*, **29**(4), 1165–1188.

Berrington de Gonzalez, A. and Cox, D. (2007). Interpretation of interaction: A review. *Annals of Applied Statistics*, **1**, 371–385.

Breiman, L. (1996). Bagging predictors. *Machine Learning*, **24**, 123–140.

Breiman, L. (2001). Random forests. *Machine Learning*, **45**, 5–32.

Breiman, L. (2003). Manual—setting up, using and understanding random forests v4.0. *http://oz.berkeley.edu/users/breiman/Using_random_forests_v4.0.pdf*.

Breiman, L., Friedman, J., Olshen, R., and Stone, C. (1993). *Classification and Regression Trees*. Chapman and Hall/CRC.

Brown, P., Vannucci, M., and Fearn, T. (2002). Bayes model averaging with selection of regressors. *Journal of the Royal Statistical Society, Series B: Statistical Methodology*, **64**(3), 519–536.

Bureau, A., Dupuis, J., Falls, K., Lunetta, K., Hayward, B., Keith, T., and Van Eerdewegh, P. (2005). Identifying SNPs predictive of phenotype using random forests. *Genetic Epidemiology*, **28**, 171–182.

Cann, H., de Toma, C., Cazes, L., *et al.* (2002). A human genome diversity cell line panel. *Science*, **296**(5566), 261–262.

Casella, G. and Berger, R. (2002). *Statistical Inference*. Duxbury Press.

Chapman, J., Cooper, J., Todd, J., and Clayton, D. (2003). Detecting disease associations due to linkage disequilibrium using haplotype tags: A class of tests and the determinants of statistical power. *Human Heredity*, **56**, 18–31.

Cheverud, J. (2001). A simple correction for multiple comparisons in interval mapping genome scans. *Heredity*, **87**, 52–58.

Chipman, H., George, E., and McCulloch, R. (1998). Bayesian CART model search (C/R: P948-960). *Journal of the American Statistical Association*, **93**, 935–948.

Chipman, H., George, E., and McCulloch, R. (2008). BART: Bayesian additive regression trees. *arXiv.org*, **stat**, arXiv:0806.3286.

Christenfeld, N., Sloan, R., Carroll, D., and Greenland, S. (2004). Risk factors, confounding and the illusion of statistical control. *Psychosomatic Medicine*, **66**, 868–875.

Christensen, R. (2002). *Plane Answers to Complex Questions: The Theory of Linear Models*. Springer-Verlag.

Christianini, N. and Shawe-Taylor, J. (2000). *Support Vector Machines*. Cambridge University Press.

Chu, G., Narasimhan, B., Tibshirani, R., and Tusher, V. (2008). SAM "Significance Analysis of Microarrays": Users guide and technical document. *http://www-stat.stanford.edu/~tibs/SAM/*, **Technical Report**, 1–41.

Clayton, D. and Leung, H. (2007). An R package for analysis of whole-genome association studies. *Human Heredity*, **64**, 45–51.

Clayton, D., Chapman, J., and Cooper, J. (2004). Use of unphased multilocus genotype data in indirect association studies. *Genetic Epidemiology*, **27**, 415–428.

Cole, S. and Hernan, M. (2002). Fallibility in estimating direct effects. *International Journal of Epidemiology*, **31**, 163–165.

Cordell, H. (2002). Epistasis: What it means, what it doesn't mean, and statistical methods to detect it in humans. *Human Molecular Genetics*, **11**(20), 2463–2468.

Costello, T., Falk, C., and Ye, K. (2003). Data mining and computationally intensive methods: Summary of group 7 contributions to genetic analysis workshop 13. *Genetic Epidemiology*, **25**(Supplement 1), S57–S63.

Cupples, L., Bailey, J., Cartier, K., Falk, C., Liu, K., Ye, Y., Yu, R., Zhang, H., and Zhao, H. (2005). Data mining. *Genetic Epidemiology*, **29**(Supplement 1), S103–S109.

Dalgaard, P. (2002). *Introductory Statistics with R*. Springer Verlag.

Demidenko, E. (2004). *Mixed Models: Thoery and Applications*. John Wiley & Sons.

Dempster, A., Laird, N., and Rubin, D. (1977). Maximum likelihood from incomplete data via the EM algorithm (C/R: P22-37). *Journal of the Royal Statistical Society, Series B: Methodological*, **39**, 1–22.

Devlin, B. and Risch, N. (1995). A comparison of linkage disequilibrium measures for fine-scale mapping. *Genomics*, **29**, 311–322.

Devlin, B. and Roeder, K. (1999). Genomic control for association studies. *Biometrics*, **55**(4), 997–1004.

Diggle, P., Liang, K.-Y., and Zeger, S. (1994). *Analysis of Longitudinal Data*. Oxford University Press.

Dudoit, S. and van der Laan, M. (2008). *Multiple Testing Procedures with Applications to Genomics*. Springer.

Dudoit, S., Shaffer, J., and Boldrick, J. (2003). Multiple hypothesis testing in microarray experiments. *Statistical Science*, **18**(1), 71–103.

Emigh, T. (1980). A comparison of tests for Hardy-Weinberg equilibrium. *Biometrics*, **36**, 627–642.

Epstein, M., Allen, A., and Satten, G. (2007). A simple and improved correction for population stratification in case–control studies. *American Journal of Human Genetics*, **80**, 921–930.

Ewens, W. and Grant, G. (2006). *Statistical Methods in Bioinformatics: An Introduction*. Springer-Verlag.

Excoffier, L. and Slatkin, M. (1995). Maximum-likelihood estimation of molecular haplotype frequencies in diploid population. *Molecular Biology and Evolution*, **12**(5), 921–927.

Faraway, J. (2005). *Linear Models with R*. Chapman & Hall/CRC.

Fitzmaurice, G., Laird, N., and Ware, J. (2004). *Applied Longitudinal Analysis*. John Wiley & Sons.

Foulkes, A., De Gruttola, V., and Hertogs, K. (2004). Combining genotype groups and recursive partitioning: An application to human immunodeficiency virus type 1 genetics data. *Journal of the Royal Statistical Society, Series C: Applied Statistics*, **53**(2), 311–323.

Foulkes, A., Reilly, M., Zhou, L., Wolfe, M., and Rader, D. (2005). Mixed modelling to characterize genotype–phenotype associations. *Statistics in Medicine*, **24**(5), 775–789.

Foulkes, A., Yucel, R., and Li, X. (2008). A likelihood-based approach to mixed modeling with ambiguity in cluster identifiers. *Biostatistics*, **9**(4), 635–657.

Free Software Foundation, I. (2007). GNU General Public License. *http://www.gnu.org/licenses/gpl.html*.

Freund, Y. (1990). Boosting a weak learning algorithm by majority. In M. Fulk and J. Case, editors, *Proceedings of the Third Annual Workshop on Computational Learning Theory*, pages 202–216. Morgan Kaufman Publishers Inc.

Freund, Y. and Schapire, R. (1997). A decision-theoretic generalization of on-line learning and an application to boosting. *Journal of Computer and System Sciences*, **55**(1), 119–139.

Friedman, J. (1991). Multivariate adaptive regression splines (with discussion). *Annals of Statistics*, **19**(1), 1–141.

Friedman, J. and Fisher, N. (1999). Bump hunting in high dimensional data. *Statistics and Computing*, **9**, 123–143.

Gao, X., Starmer, J., and Martin, E. (2008). A multiple testing correction method for genetic association studies using correlated single nucleotide polymorphisms. *Genetic Epidemiology*, **32**, 361–369.

Gelman, A. and Meng, X. (2004). *Applied Bayesian Modeling and Causal Inference from Incomplete-Data Perspectives*. John Wiley & Sons.

Gelman, A., Carlin, J., Stern, H., and Rubin, D. (2004). *Bayesian Data Analysis*. Chapman & Hall/CRC.

Gentleman, R. (2008). *R Programming for Bioinformatics*. CRC Press.

Gentleman, R., Carey, V., Bates, D., Bolstad, B., Dettling, M., Dudoit, S., Ellis, B., Gautier, L., Ge, Y., Gentry, J., Hornik, K., Hothorn, T., Huber, W., Iacus, S., Irizarry, R., Leisch, F., Cheng, L., Maechler, M., Rossini, A., Sawitzki, G., Smith, C., Smyth, G., Tierney, L., Yang, J., and Zhang, J. (2004). Bioconductor: Open software development for computational biology and bioinformatics. *Genome Biology*, **5**(10), R80.

Gentleman, R., Carey, V., Huber, W., Irizarry, R., and Dudoit, S. (2005). *Bioinformatics and Computational Biology Solutions Ising R and Bioconductor*. Springer.

George, E. and McCulloch, R. (1993). Variable selection via Gibbs sampling. *Journal of the American Statistical Association*, **88**, 881–889.

Gillespie, J. (1998). *Population Genetics: A Concise Guide*. Johns Hopkins University Press.

Givens, G. and Hoeting, J. (2005). *Computational Statistics*. John Wiley & Sons.

Goeman, J., van de Geer, S., de Kort, F., and van Houwelingen, H. (2004). A global test for groups of genes: Testing association with a clinical outcome. *Bioinformatics*, **20**(1), 93–99.

Hastie, T., Tibshirani, R., and Friedman, J. (2001). *The Elements of Statistical Learning: Data Mining, Inference, and Prediction*. Springer-Verlag.

Hawley, M. and Kidd, K. (1995). HAPLO: A program using the EM algorithm to estimate the frequencies of multi-site haplotypes. *Journal of Heredity*, **86**, 409–411.

Helbig, M., Theus, M., and Urbanek, S. (2005). JGR: JAVA GUI for R. *Statistical Computing and Graphics*, **16**(2), 9–12.

Hernan, M., Hernandez-Diaz, S., Werler, M., and Mitchell, A. (2002). Causal knowledge as a prerequisite for confounding evaluation: An application to birth defects epidemiology. *American Journal of Epidemiology*, **155**, 176–184.

Hoggart, C., Whittaker, J., De Iorio, M., and Balding, D. (2008). Simultaneous analysis of all SNPs in genome-wide and re-sequencing association studies. *PLoS Genetics*, **4**(7), e1000130.

Hosking, L., Lumsden, S., Lewis, K., Yeo, A., McCarthy, L., Bansal, A., Riley, J., Purvis, I., and Xu, C. (2004). Detection of genotyping errors by Hardy-Weinberg equilibrium testing. *European Journal of Human Genetics*, **12**, 395–399.

Hosmer, D. and Lemeshow, S. (2000). *Applied Logistic Regression*. John Wiley & Sons.

Jirtle, R. and Skinner, M. (2007). Environmental epigenomics and disease susceptibility. *Nature Reviews Genetics*, **8**, 253–262.

Johnson, R. and Wichern, D. (2002). *Applied Multivariate Statistical Analysis*. Prentice-Hall.

Kallberg, H., Ahlbom, A., and Alfredsson, L. (2006). Calculating measures of biological interaction using R. *European Journal of Epidemiology*, **21**, 571–573.

Kennedy, G., Matsuzakie, H., Dong, S., *et al.* (2003). Large-scale genotyping of complex DNA. *Nature Biotechnology*, **21**(10), 1233–1237.

Khoury, M., Beaty, T., and Cohen, B. (1993). *Fundamentals of Genetic Epidemiology*. Oxford University Press.

Kooperberg, C. and Ruczinski, I. (2005). Identifying interacting SNPs using Monte Carlo logic regression. *Genetic Epidemiology*, **28**, 157–170.

Kooperberg, C., Ruczinski, I., LeBlanc, M., and Hsu, L. (2001). Sequence analysis using logic regression. *Genetic Epidemiology*, **21**(Suppl.1), S626–S631.

Kraft, P. and Stram, D. (2007). Re: The use of inferred haplotypes in downstream analyses. *American Journal of Human Genetics*, **81**, 863–865.

Lake, S., Lyon, H., Tantisira, K., Silverman, E., Weiss, S., Laird, N., and Schaid, D. (2003). Estimation and tests of haplotype–environment interaction when linkage phase is ambiguous. *Human Heredity*, **55**, 56–65.

Lander, E. and Schork, N. (1994). Genetic dissection of complex traits. *Science*, **265**(5181), 2037–2048.

Lange, K. (2002). *Mathematical and Statistical Methods for Genetic Analysis*. Springer-Verlag.

Lehmann, E. (1997). *Testing Statistical Hypotheses*. Springer-Verlag.

Li, J. and Ji, L. (2005). Adjusting multiple testing in multilocus analyses using the eigenvalues of a correlation matrix. *Heredity*, **95**, 221–227.

Lin, D. and Huang, B. (2007). The use of inferred haplotypes in downstream analyses. *American Journal of Human Genetics*, **80**, 577–579.

Lin, D. and Zeng, D. (2006). Likelihood-based inference on haplotype effects in genetic association studies. *Journal of the American Statistical Association*, **101**(473), 89–104.

Lin, S., Cutler, D., Zwick, M., and Chakvravarti, A. (2002). Haplotype inference in random population samples. *American Journal of Human Genetics*, **71**, 1129–1137.

Little, R. and Rubin, D. (2002). *Statistical Analysis with Missing Data*. John Wiley & Sons.

Liu, B. (1998). *Statistical Genomics*. CRC Press.

Long, J., Williams, R., and Urbanek, M. (1995). An E-M algorithm and testing strategy for multiple locus haplotypes. *American Journal of Human Genetics*, **56**, 799–810.

Louis, T. (1982). Finding the observed information matrix when using the EM algorithm. *Journal of the Royal Statistical Society, Series B: Methodological*, **44**, 226–233.

Lunn, D., Whittaker, J., and Best, N. (2006). A Bayesian toolkit for genetic association studies. *Genetic Epidemiology*, **30**(3), 231–247.

Lynch, M. and Walsh, B. (1998). *Genetics and Analysis of Quantitative Traits*. Sinauer Associates, Inc.

McCulloch, C. and Searle, S. (2001). *Generalized, Linear, and Mixed Models*. John Wiley & Sons.

McLachlan, G. and Krishnan, T. (1997). *The EM Algorithm and Extensions*. John Wiley & Sons.

McLachlan, G., Do, K., and Ambroise, C. (2004). *Analyzing Microarray Gene Expression Data*. John Wiley & Sons.

Meilijson, I. (1989). A fast improvement to the EM algorithm on its own terms. *Journal of the Royal Statistical Society, Series B: Methodological*, **51**, 127–138.

Nelson, M., Kardia, S., Ferrell, R., and Sing, C. (2001). A combinatorial partitioning method to identify multilocus genotypic partitions that predict quantitative trait variation. *Genome Research*, **11**, 458–470.

Neter, J., Wasserman, W., and Kutner, M. (1996). *Applied Linear Statistical Models: Regression, Analysis of Variance, and Experimental Designs*. Richard D. Irwin.

Niu, T., Qin, Z., Xu, X., and Liu, J. (2002). Bayesian haplotype inference for multiple linked single-nucleotide polymorphisms. *American Journal of Human Genetics*, **70**, 157–169.

Nonyane, B. and Foulkes, A. (2007). Multiple imputation and random forests (MIRF) for unobservable, high-dimensional data. *International Journal of Biostatistics*, **3**(1), Article 12.

Nonyane, B. and Foulkes, A. (2008). Application of two machine learning algorithms to genetic association studies in the presence of covariates. *BMC Genetics*, **9**, 71.

Nyholt, D. (2004). A simple correction for multiple testing for single-nucleotide polymorphisms in linkage disequilibrium with each other. *American Journal of Human Genetics*, **74**, 765–769.

Pagano, M. and Gauvreau, K. (2001). *Principles of Biostatistics*. Duxbury Press.

Parmigiani, G., Garrett, E., Irizarry, R., and Zeger, S. e. (2003). *The Analysis of Gene Expression Data: Methods and Software*. Springer-Verlag.

Pearl, J. (2000). *Causality: Models, Reasoning, and Inference*. Cambridge University Press.

Pinheiro, J. and Bates, D. (2000). *Mixed-Effects Models in S and S-PLUS*. Springer-Verlag.

Pollard, K. and van der Laan, M. (2004). Choice of a null distribution in resampling-based multiple testing. *Journal of Statistical Planning and Inference*, **125**, 85–100.

Price, A., Patterson, N., Plenge, R., Weinblatt, M., Shadick, N., and Reich, D. (2006). Principal components analysis. *Nature Genetics*, **38**, 904–909.

Pritchard, J., Stephens, M., Rosenberg, N., and Donnelly, P. (2000). Association mapping in structured populations. *American Journal of Human Genetics*, **67**, 170–181.

Purcell, S., Neale, B., Todd-Brown, K., Thomas, L., Ferreira, M., Bender, D., Maller, J., Sklar, P., de Bakker, P., Daly, M., and Sham, P. (2007). PLINK: A tool set for whole-genome association and population-based linkage analyses. *American Journal of Human Genetics*, **81**(3), 559–575.

Rabbee, N. and Speed, T. (2006). A genotype calling algorithm for affymetrix SNP arrays. *Bioinformatics*, **22**(1), 7–12.

Ritchie, M., Hahn, L., Roodi, N., Bailey, R., Dupont, W., Parl, F., and Moore, J. (2001). Multifactor-dimensionality reduction reveals high-order interactions among estrogen-metabolism genes in sporadic breast cancer. *American Journal of Human Genetics*, **69**(1), 138–147.

Robertson, D., Hahn, B., and Sharp, P. (1995). Recombination in AIDS viruses. *Journal of Molecular Evolution*, **40**(3), 249–59.

Robins, J. and Greenland, S. (1992). Identifiability and exchangeability for direct and indirect effects. *Epidemiology*, **3**, 143–155.

Rosner, B. (2006). *Fundamentals of Biostatistics*. Duxbury Press.

Rothman, K. and Greenland, S. (1998). *Modern Epidemiology*. Little, Brown and Co.

Ruczinski, I., Kooperberg, C., and LeBlanc, M. (2003). Logic regression. *Journal of Computational and Graphical Statistics*, **12**, 475–511.

Ruczinski, I., Kooperberg, C., and LeBlanc, M. (2004). Exploring interactions in high dimensional genomic data: An overview of logic regression. *Journal of Multivariate Analysis*, **90**, 178–195.

Salyakina, D., Seaman, S., Browning, B., Dudbridge, F., and Muller-Myhsok, B. (2005). Evaluation of Nyholt's procedure for multiple testing correction. *Human Heredity*, **60**, 19–25.

Sasieni, P. (1997). From genotypes to genes: Doubling the sample size. *Biometrics*, **53**, 1253–1261.

Schaid, D. (2004). Linkage disequilibrium testing when linkage phase is unknown. *Genetics*, **166**, 505–512.

Schaid, D., Rowland, C., Tines, D., Jacobson, R., and Poland, G. (2003). Score tests for association between traits and haplotypes when linkage phase is ambiguous. *Human Heredity*, **55**, 56–65.

Schapire, R. (1990). Strength of weak learnability. *Journal of Machine Learning*, **5**, 197–227.

Scheet, P. and Stephens, M. (2006). A fast and flexible statistical model for large-scale population genotype data: Applications to inferring missing genotypes and haplotypic phase. *American Journal of Human Genetics*, **78**(4), 629–644.

Scheffe, H. (1999). *The Analysis of Variance*. John Wiley & Sons.

Schumacher, F. and Kraft, P. (2007). A Bayesian latent class analysis for whole-genome association analyses: An illustration using the gaw15 simulated rheumatoid arthritis dense scan data. *BMC Proceedings*, **1**(Suppl.1), S112.

Schwender, H. and Ickstadt, K. (2008a). Identification of SNP interactions using logic regression. *Biostatistics*, **9**(1), 187–198.

Schwender, H. and Ickstadt, K. (2008b). Quantifying the importance of genotypes and sets of single nucleotide polymorphisms for prediction in association studies. Technical report, Dortmund University of Technology.

Segal, M., Barbour, J., and Grant, R. (2004). Relating HIV-1 sequence variation to replication capacity via trees and forests. *Statistical Applications in Genetics and Molecular Biology*, **3**(1), Article 2.

Siegmund, D. and Yakir, B. (2007). *The Statistics of Gene Mapping*. Springer.

Spector, P. (2008). *Data Manipulation with R*. Springer.

Speed, T. (2003). *Statistical Analysis of Gene Expression Microarray Data*. CRC Press.

Stephens, M. and Donnelly, P. (2000). Inference in molecular population genetics. *Journal of the Royal Statistical Society, Series B*, **62**(4), 605–655.

Stephens, M. and Donnelly, P. (2003). A comparison of methods for haplotype reconstruction from population genotype data. *American Journal of Human Genetics*, **73**, 1162–1169.

Stephens, M., Smith, N., and Donnelly, P. (2001). A new statistical method for haplotype reconstruction from population data. *American Journal of Human Genetics*, **68**, 978–989.

Storey, J. (2002). A direct approach to false discovery rates. *Journal of the Royal Statistical Society, Series B: Statistical Methodology*, **64**(3), 479–498.

Storey, J. (2003). The positive false discovery rate: A Bayesian interpretation and the $q$-value. *The Annals of Statistics*, **31**(6), 2013–2035.

Storey, J. and Tibshirani, R. (2003). Statistical significance for genomewide studies. *Proceedings of the National Academy of Sciences*, **100**(16), 9440–9445.

Strobl, C., Boulesteix, A., Zeileis, A., and Hothorn, T. (2007). Bias in random forest variable importance measures: Illustrations, sources and a solution. *BMC Bioinformatics*, **8**(25), doi:10.1186/1471–2105–8–25.

Thomas, D. (2004). *Statistical Methods in Genetic Epidemiology*. Oxford University Press.

Thompson, P., Moyna, N., Seip, R., Clarkson, P., Angelopoulos, T., Gordon, -P., Pescatello, L., Visich, P., Zoeller, R., Devaney, J., Gordish, H., Bilbie, S., and Hoffman, E. (2004). Functional polymorphisms associated with human muscle size and strength. *Medicine and Science in Sports and Exercise*, **36**(7), 1132–1139.

Tibshirani, R. (1996). Regression shrinkage and selection via the lasso. *Journal of the Royal Statistical Society, Series B: Methodological*, **58**, 267–288.

Tukey, J. (1977). *Exploratory Data Analysis*. Addison-Wesley.

Tusher, V., Tibshirani, R., and Chu, G. (2001). Significance analysis of microarrays applied to the ionizing radiation response. *Proceedings of the National Academy of Sciences*, **98**(9), 5116–5121.

Tzeng, J., Wang, C., Kao, J., and Hsiao, C. (2006). Regression-based association analysis with clustered haplotypes through use of genotypes. *American Journal of Human Genetics*, **78**, 231–242.

van der Laan, M. (2006). Statistical inference for variable importance. *The International Journal of Biostatistics*, **2**(1), Article 2.

Vander, A., Sherman, J., and Luciano, D. (1994). *Human Physiology*. McGraw-Hill.

Venables, W. and Smith, D. (2008). An introduction to R, version 2.7.1. Technical report, The Comprehensive R Archive Network (CRAN).

Verbeke, G. and Molenberghs, G. (2000). *Linear Mixed Models for Longitudinal Data*. Springer-Verlag.

Vonesh, E. and Chinchilli, V. (1997). *Linear and Nonlinear Models for the Analysis of Repeated Measurements*. Marcel Dekker.

Wahlund, S. (1928). Zusammensetzung von population und korrelationserscheinung vom standpunkt der vererbungslehre aus betrachtet. *Hereditas*, **11**, 65–106.

Weir, B. (1996). *Genetic Data Analysis II: Methods for Discrete Population Genetic Data*. Sinauer Associates.

West, M. (2003). Bayes factor regression models in the "Large $p$, Small $n$" paradigm. *In: Bayesian Statistics 7, Clarendon Press*, pages 733–742.

Westfall, P. and Young, S. (1993). *Resampling-Based Multiple Testing: Examples and Methods for P-value Adjustment*. John Wiley & Sons.

Wittke-Thompson, J., Pluzhnikov, A., and Cox, N. (2005). Rational inferences about departures from Hardy-Weinberg equilibrium. *American Journal of Human Genetics*, **76**(6), 967–986.

Zaykin, D., Westfall, P., Young, S., Karnoub, M., Wagner, M., and Ehm, M. (2001). Testing association of statistically inferred haplotypes with discrete and continuous traits in samples of unrelated individuals. *Human Heredity*, **53**, 79–91.

Zhang, H. and Singer, B. (1999). *Recursive Partitioning in the Health Sciences*. Springer.

Ziegler, A. and Koenig, I. (2007). *A Statistical Approach to Genetic Epidemiology*. Wiley-VCH.

Zou, H. and Hastie, T. (2005). Regularization and variable selection via the elastic net. *Journal of the Royal Statistical Society, Series B: Statistical Methodology*, **67**(2), 301–320.

# Glossary of Terms

admixed population: population in which mating occurs between subgroups with different allelic distributions, or more loosely a population in which multiple subgroups are present.

allele: sequence of one or more bases, representing one of several possible forms for a region of DNA.

allelic phase: alignment of nucleotides on a single homolog.

amino acid: building block for proteins, corresponding to three adjacent bases.

biallelic: detectable presence of two alleles across a population.

Boolean expression: a logical function of multiple binary predictor variables that takes on the values 0 and 1.

candidate gene study: an investigation involving multiple SNPs within and across genes that uses information on linkage disequilibrium (LD) blocks.

candidate polymorphism study: investigation of genotype–trait association for which there is an *a priori* hypothesis about functionality.

conditional association: setting in which the effect of a predictor on the outcome is statistically significant within at least one level of a third variable.

confounder: variable associated with both a potential predictor variable and the dependent variable.

consensus: amino acid at a given site on the viral genome that is most common in the general population.

conserved: region of DNA with no observed variability within a population.

contrast: defined for the one-way analysis of variance setting as a linear combination of the means such that the coefficients sum to zero.

diplotype: pair of haplotypes, one inherited from each parental genome.

dominant: describes a type of polymorphism that results in an alteration in a trait regardless of the value of the second element of the genotype pair.

effect mediator: a variable that lies in the causal pathway between the predictor and outcome of interest; used interchangeably with causal pathway variable.

effect modification: situation in which the effect of a predictor variable on the outcome depends on the level of a third variable; used interchangeably with statistical interaction.

eigenvalue: a scalar characteristic of a matrix that is defined as the amount by which a vector in the direction of the corresponding eigenvector is stretched or shrunken when acted upon by this matrix.

eigenvector: corresponding to a matrix, a vector that spans the space that, when acted upon by this matrix is not rotated.

false discovery rate: expected proportion of null hypotheses that are true among those that are declared significant; abbreviated FDR.

family-wise error under the complete null: probability of rejecting one or more null hypotheses given that they are all true; abbreviated FWEC.

family-wise error: probability of making at least one type-1 error.

fine mapping study: investigation aimed at identifying, with a high level of precision, the location of a disease-causing variant.

functional: a polymorphism affecting a trait through a direct causal relationship.

gene expression study: investigation of the relationship between gene products, such as ribonucleic acid (RNA) or proteins, and disease outcomes or measures of disease progression.

gene: region of DNA that is eventually made into protein or is involved in the regulation of transcription.

genetic model: a model that describes the biological interaction between alleles on homologous chromosomes; includes additive, dominant and recessive.

genome-wide association study: exploratory investigation of genotype–trait association that involves characterization of a large segment (500–1000 Kb region) of DNA; abbreviated GWAS.

genotype: observed pair of DNA bases, one inherited from each parent, at a site on the genome, represented by a categorical variable that takes on values from a predefined set of discrete characters.

genotyping error: a deviation between the true underlying genotype and the genotype that is observed through the application of a sequencing method.

haplotype tagging SNPs: sites on the genome that capture overall variability within the gene under consideration and are potentially associated with disease-causing variants.

haplotype: specific combination of alleles that are in *alignment* on a single homolog and tend to be inherited together.

Hardy-Weinberg equilibrium: state in which allele frequencies are constant within a population over generations, or equivalently independence of alleles at a single site between two homologous chromosomes, also referred to as random mating; abbreviated HWE.

heterozygous: characteristic of a genotype in which the two observed bases are different, one variant allele and one wildtype allele.

homolog (homologue): one member of a pair of homologous chromosomes.

homologous chromosomes: chromosomes with potentially different alleles that carry information on the same trait or feature.

homozygous: characteristic of a genotype in which the two observed bases are the same, both variant allele or both major widtype.

honest estimate: used in the context of classification and regression trees to refer to a measure of error that will apply broadly to any sample taken from the general population.

identical-by-descent: alleles that are derived from the same ancestor.

identical-by-state: alleles with the same DNA composition that may or may not derive from the same ancestor.

*in cis*: characteristic of two alleles indicating they are on the same homolog.

*in trans*: characteristic of two alleles indicating they are on opposite homologous chromosomes.

integrase: enzyme involved in splicing viral DNA into host cell DNA.

interaction: defined statistically as the situation in which the presence of some polymorphisms alters the effects of other polymorphisms; used interchangeably with effect modification.

level of test: probability of making a type-1 error, denoted $\alpha$.

linkage analysis: an approach that aims to identify the location of a specific gene on a chromosome.

linkage disequilibrium: an association in the alleles present at each of two sites on a genome; abbreviated LD.

locus: portion of the genome that encodes a single gene or the location of a single nucleotide on the genome.

loss of heterozygosity: loss of function of an allele in an individual for whom the second allele is already inactive.

major allele: most common allele in a population; used interchangeably with wildtype.

marker: proximate SNP at which the genotype tends to be associated with the genotype at the true disease-causing locus.

meiosis: process by which a germ cell that contains 46 chromosomes, consisting of one homolog from each parent cell, undergoes two cell divisions, resulting in daughter cells with 23 chromosomes each.

minor allele: less common allele in a population; used interchangeably with variant allele.

mitosis: process of cell division that results in the creation of daughter cells that carry identical copies of a complete set of 46 chromosomes.

model of association: mathematical formulation relating genotype variables to a trait; includes additive and multiplicative.

multilocus genotype: observed genotype across multiple SNPs or genes.

nucleotide: building block of DNA, consisting of a single DNA base (A, C, T or G) linked with both a sugar molecule and a phosphate.

$p$-value: probability of observing something as extreme or more extreme than the observed test statistic given that the null hypothesis is true.

penetrance: a measure of the extent to which the presence of a disease allele results in the disease phenotype.

phenocopy: characteristic of an individual who exhibits the disease phenotype but does not carry the disease allele under study.

polymorphism: genetic variant occurring in greater than 1% of a population.

polyploidy: presence of more than two homologous chromosomes.

population genetics: study of changes in the genetic composition of a population that occur over time and under evolutionary pressures.

population stratification: presence of multiple subgroups between which there is minimal mating or gene transfer; also referred to as population substructure.

population substructure: presence of multiple subgroups between which there is minimal mating or gene transfer; also referred to as population stratification.

population-based study: an investigation involving unrelated individuals that is distinguished from family-based studies, for which correlated data methods are generally required.

protease: protein cleaving enzyme involved in the life cycle of HIV; abbreviated Pr.

$q$-value: expected proportion of false positives among all features that are as extreme as or more extreme than the feature under consideration.

quasi-species: viral population within a single human host.

recessive: describes a type of polymorphism that results in an alteration in a trait only in the presence of two copies of the polymorphism.

recombination: the joining of two broken DNA strands, one from the maternal side and one from the paternal side, that occurs as parental chromosomes are passed to an offspring.

reverse transcriptase: enzyme involved in reverse transcribing viral RNA into DNA, abbreviated RT.

single-nucleotide polymorphism: variant at a single site (base pair position) on the genome; abbreviated SNP.

single-step adjustment: multiple testing procedure in which a single criterion is used to assess significance of all test statistics or corresponding $p$-values.

statistical independence: when the joint probability of two events (alleles) is equal to the product of the two marginal probabilities.

step-down adjustment: multiple testing procedure that involves ordering test statistics or $p$-values and then using a potentially different criterion for each of the ordered values.

strains: genotypically distinct viruses resulting from multiple infections or quasispecies that developed over time within the host.

strong control: characteristic of multiple testing adjustment in which the family-wise error is less than or equal to $\alpha$ under all partial subsets of null hypotheses.

subset pivotality: when the distribution of test statistics is the same under any combination of true null hypotheses.

trait: a measure of disease status or disease progression.

type-1 error rate: probability of rejecting the null in favor of an alternative when in fact the null is true.

type-2 error rate: probability of not rejecting the null in favor of an alternative when in fact the null is false.

variant allele: less common allele in a population; used interchangeably with minor allele.

weak control: characteristic of multiple testing adjustment in which the family-wise error is less than or equal to $\alpha$ under the complete null.

wildtype: most common allele in a population; used interchangeably with major allele.

zygosity: comparative genetic makeup of two homologous chromosomes.

# Glossary of Select R Packages

gap: package includes several functions for the analysis of genetic data arising from both population- and family-based studies, including BFDP() for
calculating the Bayesian false discovery probability; gcontrol() and gcon-
trol2() for calculating genomic control statistics; hwe() and hwe.hardy()
for testing Hardy-Weinberg equilibrium using a $\chi^2$-test and Monte Carlo
methods, respectively; and hap() and genecounting() for haplotype reconstruction using sorting and trimming algorithms.

GenABEL: package developed specifically to address the computational challenges of genome-wide association studies and associated analyses on desk-
top computers. Functions include convert.snp.illumina() for converting
data from Illumuna/Affymetrix platform to genotypic data formatted file;
several data manipulation functions and summary functions, including
snp.data-class(), which provides the number of observed genotypes and
allelic frequencies for each SNP; and methods for quality control checking, including check.marker(), HWE.show() and ibs().

genetics: package includes several functions for genetic association studies,
including but not limited to LD() for calculating and testing pairwise
linkage disequilibrium; HWE.chiq(), HWE.exact() and HWE.test() for
testing for Hardy-Weinberg equilibrium (HWE); genotype(), homozy-
gote(), heterozygote() and allele.count() for coding genotype data; sum-
mary.genotype() and plot.genotype() for summarizing allele and genotype
frequencies; and dieseq.ci() for calculating bootstrap confidence interval
for single-marker disequilibrium.

hapassoc: package for estimating haplotype–trait association, including functions pre.hapassoc() for augmenting genotype data to include "pseudo-
individuals with all possible corresponding haplotypes and hapassoc() for
estimating haplotype effects using an EM algorithm.

haplo.ccs: package that includes the haplo.ccs() function for estimating haplotype relative risks in case–contol data by weighted logistic regression

and sandcov() for computing the sandwich variance/covariance estimates of the estimated coefficients.

haplo.stats: package for analysis involving unmeasured (ambiguous) haplotypes includes haplo.em() for EM approach to computation of haplotype frequencies; haplo.glm() for general linear model regression of quantitative or categorical trait on ambiguous haplotypes within a single gene; haplo.score() for calculating the score statistic for association between a trait and ambiguous haplotypes; and haplo.group() for estimating haplotype frequencies within each level of a grouping variable.

LDheatmap: package that includes function LDheatmap() for graphically displaying pairwise linkage disequilibrium (LD) between SNPs.

LDtests: package that includes functions for several different exact tests of linkage disequilibrium (LD) and Hardy-Weinberg equilibrium (HWE).

LogicReg: Package includes function logicreg() for implementing logic regression using simulated annealing algorithm or Monte Carlo logic regression.

mirf: package contains mirf() function for applying multiple imputation and random forests for ambiguous phase haploype–trait analysis and sepGeno() function for converting genotype matrix into an object with two columns for each SNP.

qvalue: package includes function qvalue() for estimating $q$-values based on a list of p-values.

randomForest: package contains randomForest() function to implement the random forest algorithm, several additional functions for summarizing and visualizing results and rfImpute() for handling missing genotype data.

rpart: package includes several functions related to recursive partitioning methods, including rpart() for fitting a classification or regression tree and prune.rpart() for applying cost-complexity pruning.

SNPassoc: package designed for whole genome-wide association studies that includes several functions, including WGassociation() for estimating and testing association between SNPs and a binary or quantitative trait under several assumed genetic models; Bonferroni.sig() for determining which SNPs are statistically significantly associated with the trait after performing a Bonferroni adjusment; GenomicControl() to give corrected $p$-values using genomic controls; LD() for computing pairwise linkage disequilibrium; inheritance() for recoding SNPs based on a genetic model; interactionPval() for calculating p-values for all pairwise SNP–SNP interactions after covariate adjustment using a likelihood ratio testing approach; and tableHWE() for testing Hardy-Weinberg equilibrium.

snpMatrix: package in the Bioconductor suite that is designed for the analysis of data from whole genome-wide association studies with several functions, including single.snp.tests() for applying the Cochran-Armitage trend test and Pearson's $\chi^2$-test; snp.rhs.tests() for fitting a generalized linear model with SNPs as predictors (independent variables); snp.lhs.tests() for fitting a generalized linear model with SNPs as the outcomes (dependent variables); ld.snp() for estimating linkage disequilibrium, including $D'$ and $r^2$; and snp.imputation() for imputing missing genotype data based on a training dataset for which SNP data are complete.

tree: package includes the tree() function for fitting a classification or regression tree and prune.tree() for cost-complexity pruning.

# Subject Index

acquired immunodeficiency syndrome (AIDS), 11, 16
AdaBoost, 211
additive model, 51, 57, 61, 62
admixture, *see* population admixture
alignment, 9
allelic phase, 6, 9, 58
amino acid (AA), 17
analysis of variance (ANOVA), 44, 46, 105, 125
anti-retroviral therapy (ART), 16
assignment operator, 216
association, *see* model of association
autosome, 14

bagging, *see* Bootstrap AGGregatING
base pairs, 17
Bayesian haplotype reconstruction, 137–140
Bayesian variable selection, 181, 209–211
Benjamini and Hochberg (B-H) adjustment, 109–111
Benjamini and Yekutieli (B-Y) adjustment, 111–112
biallelic, 4, 30, 31, 58
binary trait, 11
Bonferroni adjustment, 102–105, 110, 111, 123, 124
Boolean expression, 198
boosting, 211
Bootstrap AGGregatING, 211
bump hunting, 211

candidate gene, 2, 8, 65, 66
candidate polymorphism studies, 2
causal pathway, 12, 14
character variable, 216
chromatid, 14
class attribute, 216
classification and regression trees (CARTs), 157–179, 182
clustered data, 6
Cochran-Armitage (C-A) trend test, 42–43
collinearity, 58, 183
complex diseases, 13
complexity parameter, 175
computational partitioning method (CPM), 173
conditional association, 33, 35–37, 61, 172
confounding, 12, 33–35, 50, 54, 60
consensus, 23
conserved, 8
contrast, 107
correlation, 42
cost-complexity pruning, 158, 174–179
covariate, 7, 12
cross-over, 14
curse of dimensionality, 55

D-prime ($D'$), 67
daughter
  cells, 14
  nodes, 158
deoxyribonucleic acid (DNA), 2, 4, 8
dependent variable, 5

# Index of R Functions and Packages